Techniques for the Automated Optimization of HPLC Separations

Techniques for the Automated Optimization of HPLC Separations

John C. Berridge
Analytical Chemistry Department,
Pfizer Central Research,
Sandwich, Kent

A Wiley–Interscience Publication

JOHN WILEY & SONS

CHICHESTER • NEW YORK • BRISBANE • TORONTO • SINGAPORE

Library of Congress Cataloging in Publication Data:

Berridge, John C.
Techniques for the automated optimization of HPLC separations.

'A Wiley-Interscience publication.'
Includes bibliographies and index.
1. High performance liquid chromatography—Automation. I. Title.
QD79.C454B47 1986 543'.0894 85-12485

ISBN 0 471 90861 4

British Library Cataloguing in Publication Data:

Berridge, John C.
Techniques for the automated optimization of HPLC separations.
1. High performance liquid chromatography
I. Title
543'.0894 QD79.C454

ISBN 0 471 90861 4

Printed and bound in Great Britain.

To Eileen, Andrew, Richard and Rachael

Contents

Preface

The publication of a new book entitled 'Techniques for the Automated Optimization of High-Performance Liquid Chromatography Separations' heralds the start of a new era in chromatography. Liquid chromatography has now 'come of age'; most of the basic problems relating to the instrumentation and the mechanics have, by and large, been solved. Interest and effort are being focused on the improvement of efficiency, by improving sensitivity, selectivity and separation speed; these all comprise the general activity of chromatographic optimization.

The aim of this book is to draw together a selection of the techniques and procedures that are currently available to the chromatographer for the systematic optimization of liquid chromatography separations. The book has been written to provide readers with an understanding of the techniques that are currently in use such that one or more of them may be applied to a current problem. Additionally, by bringing together a wide range of approaches into a single volume, the book will enable those readers wishing to purchase equipment and/or software for automated optimization with a single reference source for comparison of the various and relative merits of the approaches available.

With the advent of the low cost microcomputer and the microprocessor-controlled chromatograph, many of these optimization techniques are available for the partially, or fully, automated optimization of separations. As is implied in the title of the book, the aim is not to consider only the methods that have been applied to fully automated optimization, but rather to consider the techniques and fundamental requirements such that new approaches with the potential for automation will also be included.

It is self-evident that a book of this size cannot cover all the topics concerned with the optimization of liquid chromatographic separations—the subject is too enormous for this. It is surely a truism that every chromatographer has his or her own approach to chromatographic optimization. Since most of the development time is spent in optimizing chromatographic selectivity, it is this area which will receive the most attention.

It should be noted that the number of scientific papers dealing with the optimization of separations is steadily growing and there is also a growing interest in this subject by commercial manufacturers. It is a relatively safe prediction that the next few years will see major breakthroughs in the field of automated optimization such that the word 'optimization' will be replaceable by the word 'development'.

Finally, I must express my indebtedness to the people who have helped me in various ways during the preparation of this book. Many thanks go to my colleagues in Pfizer Central Research: to Dr. Robin Platt and Mrs. Linda Broad for their comments on the manuscript and their help in reading the proofs and to the Research Library staff for their assistance with computer literature searching. I am also grateful to Peter Colley of Laboratory Data Control UK for his valuable comments on the manuscript. It is, however, to my family that the greatest thanks go: to my wife Eileen for all her help and to the children Andrew, Richard and Rachael, for their inexhaustible patience as I spent so many hours poring over a hot word processor with seemingly the rest of the house buried under piles of reprints.

Deal, Kent.
April 1985

John C. Berridge

1 Organization

1.1. INTRODUCTION

Chromatography—literally 'colour writing'—is defined as 'separation of substances by slow passage through or over absorbing etc. material'[1] from the Greek *khroma* (colour) and *-graphia* (writing). Whilst the aspect of slowness may well have been true when Michael S. Tswett first coined the name at the turn of the century, could that definition still be true today? Over just two decades, the theory and practice of chromatography have advanced such that multicomponent separations are now achieved in seconds. Chromatography has advanced to a highly sophisticated instrumental technique, with high-efficiency columns and sensitive detection methods, characterized by both high speed and high performance. This pinnacle of separation development, high-performance liquid chromatography (HPLC), is firmly established as a reliable and indispensable analytical tool of enormous scope and potential for the separation and quantitation of mixtures soluble in liquids. The power and capabilities of contemporary HPLC are immediately obvious on just a cursory examination of the literature. There is a wealth of excellent examples of the application of HPLC to practically every analytical problem we currently face—from food and drugs to the oil and petroleum industries, from clinical usage to applications in environmental monitoring; the applications are legion. Instrument manufacturers have been quick to respond to the growth in applications. The market place contains a wide range of systems that more than meet the requirements of most routine uses. Apparatus is available for analytical, semi-preparative and preparative use and new systems appear constantly, keeping well abreast of the rapid technological developments that continue apace in this fast-moving subject.

The fundamental question that must surely spring to mind is, however, how many of the published separations were truly optimized? What part in their development did serendipity take? Certainly it is fun to play with the chromatographic hardware to produce higher efficiencies and new detection possibilities. Almost unbelievable plate numbers are possible today, provided one can wait long enough for the peaks to appear. Nevertheless, in spite of the intense effort that there has been in the development of the chromatographic hardware, there will almost always still be the opportunity to improve further any given separation. Such an improvement, or optimization, of the separation will be of great benefit to the user in terms of time, or costs, and hopefully both! How many of us have been faced with the simple problem of deciding how to set about developing a new separation? How many separations have you developed which give chromatograms containing bunches of unresolved peaks separated by long, flat baselines? How does one set about spreading the peaks of interest more evenly through the separation? How can a systematic search for the best set of conditions for the sample be carried out? Furthermore, in this age of the computer, is it possible to carry out the optimization and possibly the development of a new separation entirely under computer control? What is to follow will provide some of the answers.

It is the initial development of an HPLC separation which is often the most difficult and time-consuming task that faces the chromatographer. Having made such an investment in a separation, it is unfortunate that so many argue against further optimization on the grounds of 'lack of time'. Yet in reality, if only a little thought had been given to the systematic development and optimization of the separation in the first place, dramatic savings in time or costs would have ensued. Thus it is the search for efficient optimization procedures that is now becoming of fundamental importance to HPLC as it becomes a generally accepted tool in routine use.

Almost any aspect of a chromatographic separation is amenable to optimization. For many years, most attention was focused on the design of the column and on the mobile phase flow rate.[2–6] This partially resulted from the fact that large diameter particles were in use (i.e. larger than 10 μm) and much was still to be learned about the mechanics of chromatography under high pressure and methods of preparing suitable columns. Today many of these basic problems have been overcome, although there is naturally still much interest in improving the performance of the chromatographic hardware.[7–13] More is known about the influence of solvent composition on retention, selectivity and peak shape and greater attention is being paid to the optimization of the mobile phase composition to achieve the desired separation. Further, there has been an extraordinary explosion in the availability of microcomputer-controlled chromatographic equipment, equipment allowing the programmed changing of solvent composition between successive runs. These changes can either be according to some pre-specified experimental design or on the basis of the data being generated during the experimental procedure. Perhaps a combination of the two may be used. An optimum separation can be determined either by the analyst or by the computer on the basis of some optimization algorithm or separation quality measurement.

The objectives of optimization in chromatography, stated generally, must be to find the set of conditions required to achieve the best separation in a given situation. But, exactly what is meant by *best*? A specific optimization definition for HPLC is hard to find. Definitions of the goals of HPLC optimization that might be considered include the following:[14,15]

to solve the problem with the minimum cost;
to achieve a separation goal with the minimum of time and effort;
to produce the 'best' separation possible of a given sample;
to select the mobile phase and column combination that gives baseline separation in a given time;
to achieve an optimum combination of speed of elution, sample size and resolution of solutes;
to locate the highest criterion value in the experimental region.

None of these definitions is truly adequate. The definition of the optimization goal will be dependent on the problem at hand. Indeed, one of the first steps in the optimization process must be that of defining the objectives. Without a well defined objective, no optimization will be possible.

The success of any separation depends essentially on the correct choice of the chromatographic mode to be used and the combination of the chromatographic stationary phase with the mobile phase. Instrumental aspects, however, must not be ignored. The ultimate acceptance of the separation will depend heavily on column and detector performance and the capabilities of the pumping system. Also important will be any limitations or advantages that accrue from the integrator or data system that may be added to the chromatograph. It is not intended to deal at any great length with the optimization of instrumental or data handling aspects of chromatography except inasmuch as they directly affect the separation quality. These aspects have been comprehensively considered in a number of excellent books and papers.[7–13,16] By and large, the selection of the mobile phase composition provides the primary objective. Of course, mobile phase selection cannot be considered in isolation since parameters such as flow rate and separation temperature will influence the final choice of the mobile phase composition.

Just within the immediate sphere of the separation there is a large number of parameters that can be optimized. Figure 1.1 gives an indication of the complexity of the system under consideration. The outer circle depicts the critical parameters likely to be considered in an optimization: these include factors such as the choice of column stationary phase (e.g. the carbon loading in reversed-phase chromatography), the separation temperature, the solvent strength and the amount of any modifier that may be present, the eluent flow rate (or, more correctly, its linear velocity), the length of the column and the diameter of the stationary phase particles. The inner circle indicates the essential factors which will be affected as a result of any parameter changes: these are resolution, selectivity, sensitivity and the pressure drop across the column. Between these two circles is a further level of interactions. Factors such as the solvent viscosity,

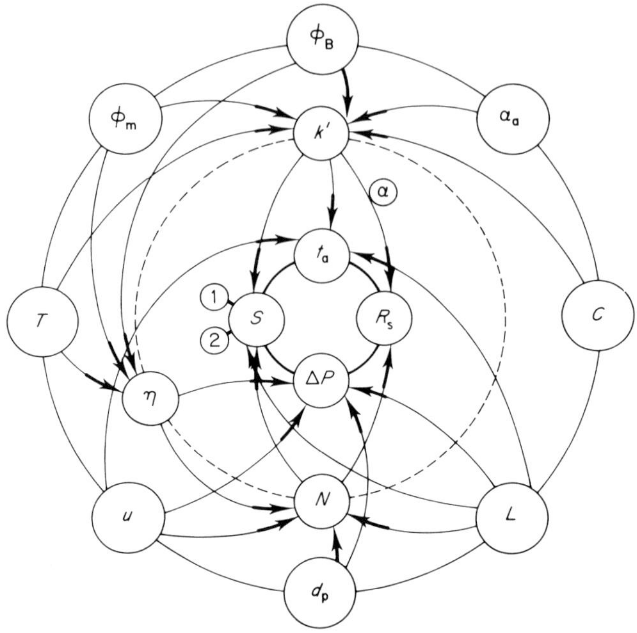

Figure 1.1. Interactions between variables (outer circle), intermediate parameters (dotted circle) and criteria of the quality of a separation (inner circle). Variables: ϕ_B = volume fraction of strongest solvent; ϕ_m = volume fraction of a modifying solvent; T = temperature; u = (linear) flow velocity; d_p = particle size; L = column length; α_a = adsorbent activity; C = carbon content of alkyl-bonded silica, or counter-ion concentration. Intermediate parameters: k' = capacity factor; N = column efficiency; η = solvent viscosity; α = selectivity factor. Criteria of quality: t_a = analysis time; R_s = resolution; P = pressure drop; S = response factor (peak height) for low-concentration solute available either in a large amount (1) or in a limited amount (2). *Reproduced from ref. 17 by permission of Elsevier Science Publishers BV*

the column efficiency and the capacity factors of the solutes are determined by the constraints imposed by the critical parameters detailed on the outer circle. These intermediate factors might almost be considered the link between the two circles: they mediate between the two, as it were translating the effects of the critical parameters to the overall separation quality. In Figure 1.1, the major interactions are highlighted by arrows.

It is the interactions and interdependence of the parameters that give rise to the complexities of HPLC optimization. However, this is not the whole picture, since there is still the problem of how a 'good' separation may be defined and recognized.

1.2. THE BASIC PROBLEM

It has already been seen that a definitive optimization objective is going to depend on the current problem. A generic criterion might be—the adequate separation of the solutes of interest within a reasonable analysis time and at an acceptable cost. Good resolution between solutes is possible in column liquid chromatography only if there is adequate control over the differential migration of the solutes (i.e. the separation selectivity) and over the column efficiency, which will be governed by the extent of the dispersion of the zones for each of the solutes as they progress through the column. These considerations are common to all forms of chromatography. Irrespective of the mode of separation being employed (e.g. ion-exchange, reversed-phase or normal-phase), the selectivity achieved between solutes will arise from differences between the distribution equilibria for the individual components between the stationary phase and the mobile phase. The dispersion of zones originates from many causes, including multiple flow paths, voids in the column or in coupling units, slow mass transfer to and from the stationary phase and axial diffusion in the column. The ultimate picture that emerges at the end of the column is thus a combination of the kinetic effects, shown as the efficiency of the separation, and the thermodynamic effects, expressed as the solute capacity factor. Combining these two effects allows a description of the separation of peaks on a chromatogram, usually expressed in terms of the resolution R, where R can be related to the variables controlling the separation as shown in equation 1.1:[8]

$$R = \tfrac{1}{4}(\alpha - 1)N^{1/2}k'/(1 + k') \tag{1.1}$$

where k' is the average value of the capacity factors of two adjacent peaks, α is the selectivity (k'_1/k'_2) and N is the column plate number. The capacity factor, k', is defined as

$$k' = (t - t_0)/t_0 \tag{1.2}$$

where t is retention time and t_0 is the column void time.

For most practical purposes, the three terms in equation 1.1 can be optimized separately, although they are not entirely independent.

The optimization of the plate number, N, that is the efficiency of the chromatographic column and hence the separation efficiency, is well described in the literature.[8-13] Manipulation of the value of the average capacity factor does not provide a great deal of scope for resolution optimization and it will be found that optimum values of k' generally lie in the range 1–10, with 4 being an ideal compromise between resolution, speed and sensitivity. It is the optimization of α, the selectivity term, that provides the greatest scope and hence often also consumes the largest proportion of time.

In contrast to gas chromatography, the mobile phase in liquid chromatography plays the dominant role in the separation process. The mobile phase composition

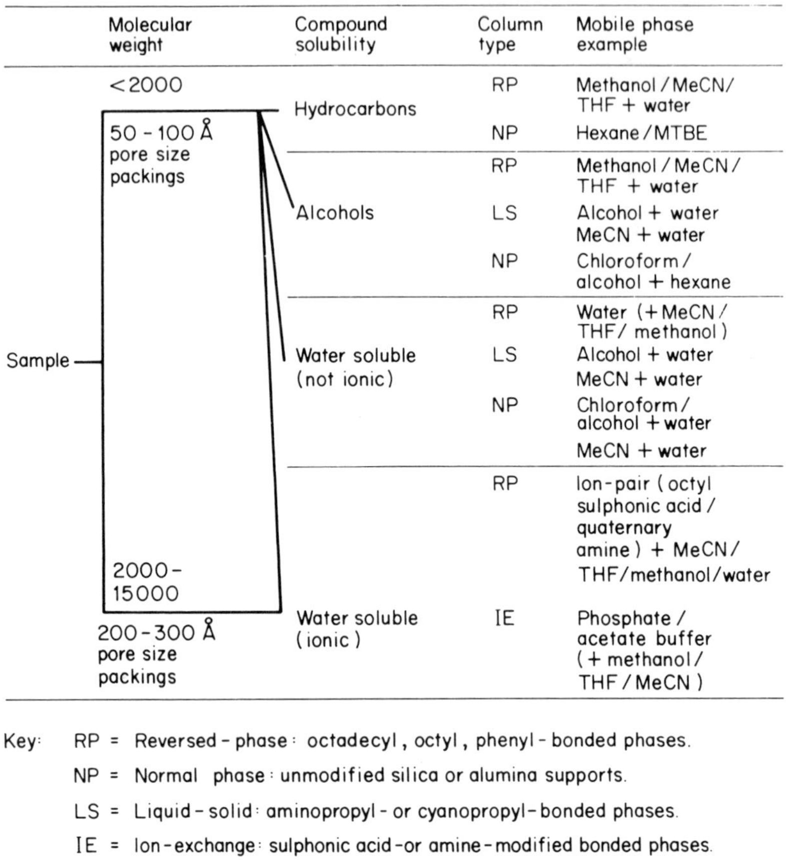

Figure 1.2. Guide to the selection of chromatographic mode, support and mobile phase type

determines the separation of the components in a mixture by controlling the degree of resolution, the selectivity of the separation (i.e. the order of elution of the peaks) and the retention times of the peaks. The selectivity and resolution of the separation will depend primarily on the specific composition of the mobile phase, whereas the retention times will be heavily dependent on the overall polarity of the mobile phase. Selection of the mobile phase in HPLC must nevertheless be based on the properties of the column stationary phase, which will have been chosen after consideration of the known properties of the solute mixture (see Figure 1.2). Even this figure is already a restriction on the possible methods which could be used. There is no mention made of dynamically coated silica for reversed-phase chromatography,[18] graphitized carbon black supports[19,20] or supercritical fluid chromatography,[21–24] to name just three further options. Unfortunately, the detailed consideration of these methods is outside the scope of this book—indeed, very little guidance is currently available on their systematic optimization.

The properties of the mixture to be separated, the equipment and the solvents available will all naturally influence the chromatographic mechanism to be used. Already a complex decision process has been initiated which will necessitate a considerable degree of expertise on the part of the chromatographer. Some help may come from the new generation of so-called 'Expert Systems'[25,26] (see also Chapter 7), but their ready availability and proven abilities are still some way off. In addition to the basic chromatographic support and mechanisms to be used as suggested by Figure 1.2, e.g. normal- or reversed-phase, ion-exchange or ion pairing, due consideration must be given to the precise technique to be employed. Thus, will isocratic or gradient separations be used? Will column switching or 'injection loading' (see Chapter 3) be considered?

The selection of mobile phase composition has traditionally been carried out largely by trial and error methods. Such methods should hardly be ignored since they have at least stood the test of time and are amply illustrated by many of the best published separations. Indeed, excellent guidance for such an approach is available in the standard texts on chromatography.[8] It is a salutory lesson to step back and appreciate the efficiency of the human computer. It is unfortunate that this otherwise excellent optimization machine can be prejudiced, or just unlucky. The desired goal may not be achieved, even after months of dedicated effort. The simple experimental approach therefore is being replaced steadily by more formal optimization strategies, which are the subject matter of this book. Many of these strategies have been, or easily could be, applied to the on-line, fully automated optimization of HPLC separations.

1.3. SCOPE AND FORMAT

The primary aim of this book is to consider techniques for the automated optimization of separations and to bring together in a single source the many that are now available. As the subject of optimization of separations has become ever more popular in its own right, an increasing number of short reviews has appeared,[27–32] but these tend to have concentrated on particular approaches to the problem. An additional problem is in the automation of the optimization step. This will mean that a small number of otherwise perfectly adequate methods will have to be omitted on the grounds that their automated implementation has not been described or is difficult for the non-expert.

Each chapter is intended to be self-contained, dealing with methods employing one particular theme, although common threads will be found running through the whole book. For example, Chapter 2 considers the initial system requirements of chromatographic hardware and optimization criteria and functions. Both of these topics are pertinent to all subsequent sections. Chapter 3 discusses some relatively simple approaches to the optimization problem. These approaches might form the starting places for more extensive and systematic methods which are described in the succeeding chapters. In Chapter 4 the very powerful statistical experimental designs, especially factorial and mixture designs, are considered and the use of overlapping resolution maps to indicate the regions of optimum separation is described. The single subject of 'window diagrams', a more general optimization method which is not restricted to chromatography, is considered in Chapter 5. Chapter 6 discusses adaptive approaches to the problem of separation optimization. The main method here is based on the sequential simplex procedure, another powerful general optimization technique which has found uses in many branches of analytical chemistry. In Chapter 7 a brief look is taken at the fundamental question of how a decision might be made on the choice of optimization method. Additionally, the possibilities of enhancing the presentation and extraction of information from a given separation by using powerful detection systems and by using various mathematical techniques is discussed. Chapter 7 also takes a speculative peek into what the future may hold. The book concludes with Appendices containing experimental details and software listings for some of the methods which have been described.

As a link between the chapters, a synthetic mixture has been subjected to the optimization procedures in order to assess their capabilities and to illustrate aspects of particular methods. There is nothing magical about the sample—it was chosen simply as a tool to enable a comparison to be made of the advantages and disadvantages of various methods. Full details concerning the mixture and the chromatographic hardware that were used for the experiments are presented in Appendix 1.

As part of each chapter, comments have been included on literature applications and on the availability of software for computer-controlled chromatographs. We must all recognize our debt to the commercial manufacturers for many of the technological advances that have given us the equipment we use today. This is, unfortunately, something of a double-edged sword since good, innovative software has commercial value. Good software is thus hard to come by, but many authors may be prepared to release their programs, particularly if there can be some understanding on confidentiality.

1.4. REFERENCES

1. *Concise Oxford Dictionary*, 6th ed., Oxford University Press, London, 1976.
2. L. R. Snyder, *J. Chromatogr. Sci.*, **10** (1972) 200.
3. L. R. Snyder, *J. Chromatogr. Sci.*, **10** (1972) 369.
4. J. H. Knox, *J. Chromatogr. Sci.*, **15** (1977) 352.
5. M. Martin, C. Eon and G. Guiochon, *J. Chromatogr.*, **99** (1974) 357.
6. L. R. Snyder, *J. Chromatogr. Sci.*, **15** (1977) 441.
7. N. A. Parris, *Instrumental Liquid Chromatography*, (Journal of Chromatography Library, Vol. 27), 2nd ed., Elsevier, Amsterdam, 1984.
8. L. R. Snyder and J. J. Kirkland, *Introduction to Modern Liquid Chromatography*, 2nd ed., Wiley, New York, 1979.
9. R. E. Kaiser, *J. High Resolut. Chromatogr. Chromatogr. Commun.*, **2** (1979) 679.
10. R. E. Kaiser and E. Oelrich, *Optimisation in HPLC*, Hüthig, Heidelberg, 1981.

11. P. Kucera (Ed.), *Microcolumn High Performance Liquid Chromatography*, Elsevier, Amsterdam, 1983.
12. R. Rosset, M. Caude, J., Desbarres and E. Schmidt, *Analusis*, **8** (1980) 213.
13. I. Halasz and G. Gorlitz, *Angew. Chem., Int. Ed. Engl.*, **21** (1982) 50.
14. Discussion Forum on Optimization Methods, 8th International Symposium on Column Liquid Chromatography, New York, 1984.
15. M. T. W. Hearn, 'HPLC of Peptides', in Cs. Horvath (Ed.), *High Performance Liquid Chromatography, Advances and Perspectives*, Vol. 3, Academic Press, New York, 1983.
16. R. E. Kaiser, *Computer Chromatography*, Hüthig, New York, 1982.
17. J. P. Bounine, G. Guiochon and H. Colin, *J. Chromatogr.*, **298** (1984) 1.
18. J. F. K. Huber, M. Pawlowska and P. Markl, *Chromatographia*, **19** (1984) 19.
19. I. Molnar and Cs. Horvath, *J. Chromatogr.*, **142** (1977) 623.
20. P. Ciccioli and R. Tappa, *J. Chromatogr.*, **269** (1983) 47.
21. P. A. Peaden and M. L. Lea, *J. Liq. Chromatogr.*, **5**, Suppl. 2 (1982) 179.
22. P. J. Schoenmakers, *J. Chromatogr.*, **315** (1984) 1.
23. M. G. Rawdon and T. A. Norris, *Int. Lab.*, **14**(5) (1984) 12.
24. K. D. Bartle, *Lab. Pract.* **34**(1) (1985) 59.
25. A/C Interface, *Anal. Chem.*, **56** (1984) 1200A.
26. J. Karnicky, *Anal. Chem.*, **56** (1984) 1312A.
27. J. C. Berridge, *Trends Anal. Chem.*, **3** (1984) 5.
28. J. Vajda and L. Leisztner, *Anal. Chem. Symp. Ser.* **16** (1984) 103.
29. S. N. Deming, J. G. Bower and K. D. Bower and H. J. Issaq, in J. C. Giddings, E. Grushka, J. Cazes and P. R. Brown (Eds.), *Advances in Chromatography*, Vol. 24, Marcel Dekker, New York, 1984, Ch. 2 and 3.
30. P. J. Colley, *Lab. Pract.*, **34** (1985) 82.
31. L. de Galan, *Trends Anal. Chem.*, **4** (1985) 62.
32. G. D'Agostino, L. Castagnetta, F. Mitchell and M. J. O'Hare, *J. Chromatogr.*, **338** (1985) 1.

2 Instrumental Requirements and Optimization Criteria

Three of the questions which should spring to mind when considering automated separation optimization should be (a) what apparatus is available, (b) what is the objective of the optimization and (c) what method(s) will be used? There is no easy answer to the first question, except to say that many of the techniques to be described can be run manually with only the simplest of apparatus to achieve a systematic optimization of a separation. However, since the aim of this book is primarily to discuss techniques suitable for automated separation optimization, this chapter begins by considering the various components that comprise a modern high-performance liquid chromatograph, with emphasis on the needs of automation. The second question needs to be answered before experimentation begins. It is essential that the objective of the work should be defined before the experiments are conducted. A favourite way of doing this for automated separation optimization is to define some expression that relates to an assessment of chromatographic performance and then to optimize the value of this expression. These expressions, or optimization criteria, are used in many of the automated optimization schemes and a review of those available forms the second part of this chapter. The final question, that of the choice of method, will be considered in subsequent chapters.

2.1. INSTRUMENTAL REQUIREMENTS

In dealing with instrumental requirements for automated separation optimization, certain tacit assumptions must be made. Thus, the merits of various individual components of a liquid chromatograph will not be discussed. Rather, a generic description of a system, or systems, that meets minimum requirements will be sought.[1] Such a system is depicted in Figure 2.1. Most modern microprocessor-controlled liquid chromatographs can carry out automated separation optimization to some degree. The basic requirements are

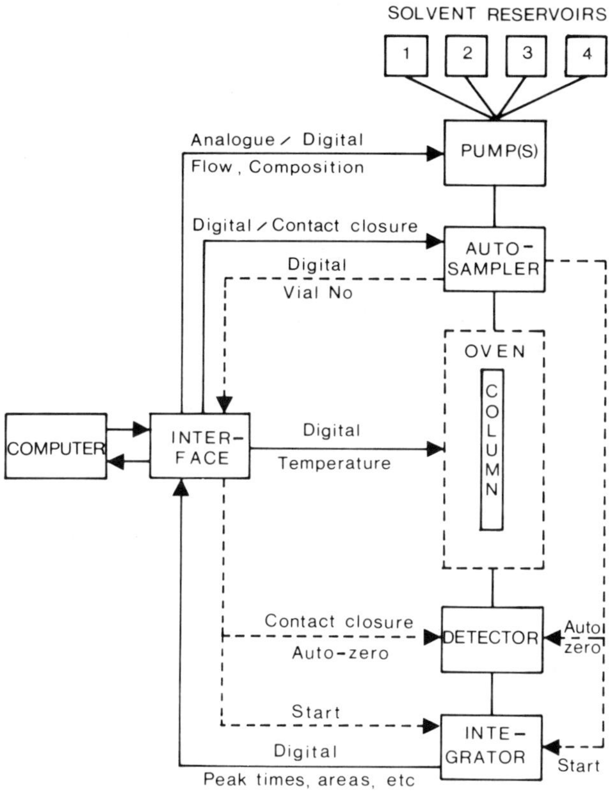

Figure 2.1. The components of a high-performance liquid chromatograph for automated separation optimization showing the essential (solid lines) and subsidiary (dashed lines) communication links

a pumping system that can be remotely controlled (in terms of the solvent composition it can deliver), a detection system suitable for the sample (e.g. UV, possibly at low wavelength), an autosampler or repeat injector and an integrator/ printer that can display the chromatograms. Ideally there will be a computer linked to the system to add a measure of artificial intelligence, but ultimately it will be the chromatographer who makes the final decision on the acceptability of a method. In considering the individual components in greater detail, attention will be focused upon components suitable for the usual range of normal- and reversed-phase chromatography. Thus equipment for supercritical fluid chromatography[2] or size exclusion separations will not be discussed.

2.1.1. Solvent Delivery

To achieve solvent composition optimization, the minimum pumping requirement is for a pump system which permits the reproducible blending of two or more solvents. There is a continuing debate on the relative merits of low-pressure versus high-pressure mixing. The advantages of low-pressure mixing, using proportioning valves and a single high-pressure pump, come from the potentially higher precision where small proportions of one or more solvents are being blended. High-pressure mixing with two or more conventional pumps cannot cope adequately with the demands of delivering just a few percent of a component. It is also easier to translate a solvent composition delivered by a low-pressure mixing system to a subsequent pre-mixed mobile phase. The problems associated with solvent compressibilities, and the volume changes ensuing from mixing, particularly methanol/acetonitrile/water mobile phases, often cause difficulties in reproducing the stated solvent composition from a high-pressure pumping system. A serious drawback with low-pressure mixing, however, is the potential for solvent degassing on mixing and subsequent failure of the high-pressure pump to deliver the mobile phase owing to pump cavitation. It is essential that adequate means are provided for purging the solvent reservoirs of dissolved air—this will almost certainly require a source of helium.

There are now solvent delivery systems available which combine the best of both low- and high-pressure mixing and thereby avoid the drawbacks of both. The individual solvents are delivered under a low, positive pressure (a few bar) by a precision volumetric pump and then boosted to the necessary high pressure by a second pump. This method has the volumetric accuracy of the low-pressure systems, and hence the translatability to a pre-mixed composition, coupled with the insensitivity of high-pressure mixing to dissolved gases in the solvents.

The number of solvents that should be available is another subject of debate. A large number of separations can be achieved with just two solvents, but few would argue about the gains in selectivity which may be achieved by the use of a third solvent.[3] The advantage gained by going to four or even more solvents is, however, not clear and it is increasingly argued that it is unnecessary to consider the optimization of four independent solvents.[4,5] To have a four-solvent delivery system available does give an extra degree of freedom and certainly provides greater flexibility where secondary solvent effects such as pH or ion-pair reagent concentration are being considered. If a four-solvent system is not available, it is possible to extend the flexibility of a two- or three-solvent system in a number of ways. The addition of a solvent selection valve to one or more of the low-pressure inlet lines to the pump can provide the opportunity to select from a number of solvents. Another way of increasing capability, particularly where prior knowledge of the separation is at hand, is to pre-blend one or more solvents.[4,6]

2.1.2. Sample Introduction

An automated sample introduction system is a prerequisite for fully automated separation optimization. There are no great constraints on the abilities of such an injector. For many applications it is not essential that the repeatability and precision should be high, although this will be necessary if peaks are being tracked on the basis of their areas, for example. It may help if some timing or control is available in the injector—a new injection must not be made until the current separation, data handling, calculation, new parameter set-up and equilibration have all been completed. There are a number of commercially available autosamplers that are eminently suitable but, if the system is going to be dedicated to separation optimization, a simple repeat sampler will be more than adequate. Such a sampler can be based on a peristaltic pump recirculating the mixture of interest through a remotely controlled injection valve. A small degree of dilution is likely to occur during the optimization sequence but this will be of little consequence.

2.1.3. Column Choice

The choice of the separation column is entirely in the hands of the chromatographer (but see Figure 1.1 first). By far the most popular mode of HPLC is the reversed-phase mode: it is estimated that over 80% of published separations have employed reversed-phase chromatography. Today's so called 'reversed-phase' separations are further complicated by the utilization of ion suppression, ion-pairing, ligand exchange and other secondary equilibria. There is also the choice of support for reversed-phase chromatography. Usually, an octadecyl-bonded support will be a good choice but there are, for example, also octyl-, phenyl-, hexyl-, ethyl- and methyl-bonded phases, all suitable for reversed-phase separations, in addition to cyano- and aminopropyl-coated supports which can be used in both normal- and reversed-phase modes. The properties of the more popular, commercially available supports have been reviewed by several authors.[7-9] Instead of using conventional reversed-phase supports, dynamically coated silica[10,11] can be considered as an alternative. Dynamically coated silica is reported to offer the advantage of a far greater selectivity reproducibility than is currently attainable with commercially available bonded supports. More recently, polymeric phases[12] have been gaining in popularity and some encouraging separations have been achieved on porous, graphitized carbonaceous materials. Tailor-made supports are also gaining in popularity for specialized separations: this is no better exemplified than with chiral stationary phases.[13,14]

It is also necessary to decide on the physical parameters of the column, its length and diameter and the diameter and nature of the support material. In general, a longer column can be expected to provide a higher theoretical plate number (but one should be aware that commercial manufacturers derive column efficiencies in a variety of ways[15]) although this does not seem to be the universal situation. Issaq[16] found that it was the first few centimetres of the column that had the dominant effect on the separation and that short columns[17] could provide separations comparable to those attained on longer columns, provided that a suitable adjustment of the mobile phase composition was made. There will be an optimum particle size for the particular separation in mind and it has been shown[18] that 1–3 μm particles are more likely to be suitable for relatively simple separations using short columns. For the more complex separations it is better to use 10–20 μm diameter particles in long columns. There is, however, an increasing trend towards shorter columns containing smaller particles because of the gains in speed which accrue, but these columns place much greater demands on the performance of the rest of the chromatographic hardware,[19] in particular the detection system.

The ultimate choice of column must be one based on experience and knowledge of the solutes. It is decisions such as this that distinguish between methods development and optimization—solvent compositions can be optimized after the decision on the column support has been made. It is important to bear in mind that a column used for automated optimization may be subjected to extremes of solvent compositions, temperature and perhaps flow rates. The chosen column should stand up to the challenges without too much deterioration. This is an area where a polymeric column may have advantages since it will not suffer from the problem of gradual dissolution that occurs with silica-based columns. For automated optimization where time is important, as short a column as is practicable is recommended, as it will allow more experiments to be conducted in a given period.[20] A short column will give greater freedom to explore different flow rates, since back-pressures will be relatively low and should not prove to be a limitation. Even if full resolution is not achieved with a short column, transfer to a column with an identical packing but longer length may provide the necessary plate number to achieve the desired separation. Alternatively, post-chromatographic mathematical techniques such as generation of derivatives or the use of peak deconvolution methods may be employed (see Chapter 7).

Another feature that is likely to be of benefit for comprehensive separation optimization is the control of column temperature. The effects of temperature on selectivity are often ignored, but may prove useful. The problems associated with temperature control usually arise from the time taken to heat or cool the column. Forced air ovens offer advantages in speed, but systems where temperature control utilizes the Peltier effect have the advantage of more rapid cooling. Where microbore columns are being used it may even be possible to vary the temperature during the separation[21] and produce temperature programming effects similar to those frequently employed in gas chromatography.

2.1.4. Detector

For any system, the minimum detection requirement is for a ‘black box’ that indicates the relative proportions (concentrations) of the components of interest and which outputs a signal proportional to these concentrations according to the time of their exit from the separation column. The fixed or variable-wavelength ultraviolet (UV) detector is likely to be the most frequently selected (currently used in about 70% of published separations[22]) but, for example, fluorescence, refractive index and electrochemical detectors should not be ignored. Using two detectors in series is a very powerful way of gaining extra information about the nature of a chromatographic peak. A very simple way to achieve this is just to run two UV detectors in series, operating at different wavelengths. The disadvantages here are the band broadening that will occur and the time delay between the two signals. There are some detectors which have the ability to monitor two or more wavelengths simultaneously, thereby eliminating these drawbacks. These detectors can also provide a variety of outputs, including difference signals or absorbance ratio signals. Where two detection methods can be coupled, for example UV with electrochemical detection, the information content will increase dramatically but, unfortunately, so will the problems of interpretation.

Whichever detector is chosen, one should aim for one which gives the highest signal-to-noise ratio and the lowest drift that can be afforded. Detector noise is a significant problem in a number of optimization schemes and can lead to false conclusions. Detector stability is also a problem. Many things will be changing during the course of an optimization. A signal that is intrinsically noisy or drifting will not help. One feature of any detector that is extremely useful is an auto-zeroing capability. When extreme solvent compositions are being investigated it is almost certain, especially if high sensitivities are being used, that the steady-state signal produced with no solute in the detector cell will vary considerably. Some integrators have great difficulties in coping with negative voltages, even of the order of a few millivolts. Large, positive signals are not generally a problem. If an auto-zero is available it should be connected to a device that can indicate when a new injection has been made: this will probably be the autosampler. If auto-zeroing is not available on the detector, then it is probably wise to set a small positive offset on the detector signal if downwards signal drift could occur. Alternatively, the zero could be set at the start of the optimization with the lowest absorbing mobile phase composition present in the detector cell.

In the future, it is likely that the so-called 'multi-channel' detectors will be increasingly used. Two excellent examples are the linear photodiode-array UV detectors that are already available and the newer multi-channel fluorescence detectors that, at the time of writing, are just emerging from the research phase. HPLC coupled to mass spectrometry is fast becoming a routine tool, providing enormous amounts of structural data on each chromatographic band. The potential for automated optimization if an on-line mass spectrometer could be used as the detector is vast. This new generation of detectors is of great interest to those pursuing automated optimization since they at last provide a means of tracking individual components in the chromatogram, thereby making separation optimization easier, quicker and more reliable. Most automated work will, however, continue to be carried out on the simpler equipment.

2.1.5. Integration

The ability to integrate the chromatogram is not a prerequisite for automated optimization. A chromatograph that can follow a pre-programmed sequence of solvent compositions can provide a large number of chromatograms automatically, which are then available for perusal by the chromatographer.

Where a degree of interactive optimization is sought, that is, interaction between the chromatograph and a digital computer, then integrated data are required. The easiest solution is to use an integrator that can communicate with the outside world via some 'standard' interface or network (see below). Many of the modern integrators have RS-232C or IEEE-488 interfaces fitted. It is possible to use microcomputers as integrators and examples of such systems are available[23] together with simple software[24]. It is likely, however, that more powerful peak integration will be required. Some have chosen to achieve this by the use of mini- or mainframe computers[25,26]. If a powerful computer controller is being used then integration may be carried out by this, either with provided software or by user-written routines.

2.1.6. Computer Control

Fully automated optimization necessarily requires some kind of computer controlling the system, and the key feature here is communication. The computer can be anything from an inexpensive microcomputer all the way up to a mainframe system. Whatever is used, however, the requirements are the same. There should be comprehensive two-way communication between the computer and the key components of the chromatograph. Should such communication channels not be available, much can still be done. Many of the techniques to be described are admirably handled by a combination of a semi-automated chromatograph and human operator. Looking again at Figure 2.1, which depicts a generalized system for automated use, the distinction is made between the communication links that are 'musts' and those that are 'nice to haves'.

The two key links are between the controlling computer and the pumping system and between the computer and the detector or integrator. The pump link is essential for full control of the mobile phase composition and most of the contemporary pumps fall into one of three categories. The simplest link and method of control is by the direct application of an analogue voltage which is proportional to the desired flow rate (e.g. Laboratory Data Control ConstaMetric and MiniMetric pumps). Also relatively straightforward are the pumps which require just application of a square-wave frequency that is proportional to the pump flow rate (e.g. Waters Associates pumps). The third category includes those pumps which have a degree of built-in intelligence such that they can process and act on instructions from other sources (e.g. Gilson pumps). The link may be specific to the manufacturer or may be through one of the standard interfaces (RS-232C or IEEE-488) or by a simple TTL (Transistor Transistor Logic) link utilizing binary coded decimal (BCD) protocols. Suitable interface cards for most popular microcomputers are readily available, particularly for the first two categories[27] and for standard communications interfaces. The only difficult part is obtaining a cable to match the component parts; acquiring this is often harder than it may appear! Fortunately, the HPLC manufacturers are increasingly building the communication links directly into the chromatographs and their controllers. Programs written by the chromatographer in high-level languages (usually BASIC or Pascal) can access the controlling files and parameters of the computer and change their values, thereby setting up new experimental conditions. The trend is for this capability to increase further and it is by far the easiest approach to achieving automated control.

The link with the detection system should be simpler to achieve but in practice often turns out to be the opposite. The detector signal will usually be a simple voltage in the range 0–10 mV or 0–1 V which is proportional to the concentration of substance being detected in the flow cell. Unfortunately, a considerable degree of processing must be done to this signal before it can be used for the input to an optimization scheme. All the optimization functions and criteria to be described rely on there being some measure of peak separation. To achieve this it will be necessary to carry out either real-time or post-run evaluation of the chromatogram. The first requirement will therefore be to capture and digitize the chromatogram. This in itself is comparatively easy with one of the standard analogue to digital interfaces that are available for most computers, except that the use of short columns packed with small-diameter particles for high-speed separations does place some severe demands on the amount of data that must be captured per second.[28] Most often integration and evaluation will be carried out post-run, which means that the digitized chromatogram must be stored on disc or in computer memory. To store a chromatogram directly in memory will require considerable amounts of random access memory (RAM) to be available unless a data compression routine, such as are frequently built in to commercial integrators, is used. A far easier way to achieve this step of data acquisition is to use a commercially available integrator. Many contemporary integrators will carry out the digitization and store the data slices in memory or on some off-line storage medium that can subsequently be accessed by the controlling computer. Having stored a digitized chromatogram it can subsequently be processed to measure the required chromatographic qualities.

The automated measurement of peak separation is not an easy task. Unless one is an extremely competent programmer, with a full knowledge of peak-smoothing algorithms and integration routines, the detection and integration of peaks is best left to the experts. Most commercially available integrators incorporate powerful integration routines designed to cope with the very worst that HPLC detectors can throw at them. The information that is available following integration will be very dependent on the particular integrator being used. The minimum information required is peak retention times, but considerable further advances can be made if peak heights and areas are available. Some integrators also make available the cardinal points, that is the positions of the start, apex and end of each peak. This information can be very useful if it is desired to consider the shape of peaks or the efficiency of a separation. The interpretation of cardinal points can be complicated where incomplete resolution occurs and reference to the integrator instruction manual will be needed to establish how best to use these data.

To transfer the integrated data to the computer, it is most likely that the 'standard' RS-232C and IEEE-488 interfaces will be required, unless a computer controller is available which incorporates a high-level language that can directly access the integrated chromatographic data. Some workers have chosen to develop their own systems[29,30] but there are a number of systems commercially available with just this capability and, as computer power increases, the number and choice of such systems will surely also increase. There is an increasing move towards the use of networks to control chromatographic systems and the use of such networks (e.g. LabNet by Spectra-Physics and DECNET by Digital Equipment Corp.) obviously provides a very easy approach to the communications problem since much of it has been taken care of by the system's manufacturer.

Access to the integrated data, that is the cardinal points or the peak retention times, heights and areas, gives considerable scope for chromatogram assessment. Minimum alpha plots (window diagrams—see Chapter 5) need only retention time data in their simplest form. It is much more likely that some estimates of resolution will be required. Resolution is defined by equation 2.1:

$$R = 2(t_2 - t_1)/(w_1 + w_2) \tag{2.1}$$

where t is the peak retention time and w is the peak width at the baseline.

The objective of resolution determination is to provide at least semi-quantitative information about the degree of separation between adjacent peaks. It is not essential for automated separation optimization that the result be absolute; it is much more important that it should be consistent and repeatable from one experiment to the next. Estimates of resolution may be made by assuming that the column plate count (N) remains approximately constant throughout the separation:[31]

$$R = [N^{1/2}/2]\,[(k'_2 - k'_1)/(2 + k'_1 + k'_2)] \tag{2.2}$$

where k' is the peak capacity factor.

It is not always possible to obtain easily the values of capacity factors or peak widths from the data provided by integrators or from standard integration routines. The assumption can perhaps be made, however, that the peaks are approximately Gaussian, in which case peak width can be estimated[32] from:

$$w = 4A/(2\pi)^{0.5}h \tag{2.3}$$

where w is the peak base width, A the peak area and h the peak height. In practice, chromatographic peaks are rarely Gaussian or symmetrical and equation 2.3 will therefore underestimate peak base width. A modification of equation 2.3 has been used for estimating peak width and hence, from equation 2.1, the peak resolution[33] in automated optimization applications:

$$w = 2A/h \tag{2.4}$$

Equation 2.4 essentially approximates the chromatographic peak to a triangle.

2.2. SOFTWARE

The question of the source and nature of the controlling software for chromatographic optimization is one with no simple answer. To deal with the nature of the software first, one is faced with the question as to whether the software should be accessible to the chromatographer for personal manipulation or whether it should be essentially transparent to the user. To a large extent these questions will be answered by the type of computer controller being employed.

Where a commercially available chromatography control system is being used, then it is highly likely that software will be provided in ROM (read-only memory) or perhaps on a set of EPROMs (erasable programmable read-only memory). To the user no difference in use will appear as neither system will allow modification of the software. The provision of the software using EPROMs is likely to be advantageous since it should be easier to acquire software updates from the supplier as it is simply a matter of reprogramming the 'chip'.

Perhaps the best method is to have software provided on a suitable magnetic medium (cassette or floppy disc), written in a high-level language, so that full access is available for the intrepid programmer to modify the program as desired. One should be aware that not all software provided on discs is accessible since it may be protected in some manner, or it could be presented in a compiled form such that comprehension and modification are impossible.

The third option open to those with suitable hardware is to 'do-it-yourself'. This method has all the advantages of gaining a complete understanding of what is going on and it also allows complete freedom in the selection of the method to be used. The major drawback to this seemingly simple approach is that of the time and resources required to achieve the objective. Having studied the chapters which follow, the reader will be in a more enlightened position as to which approach might be appropriate, but the difficulties of writing optimization software from scratch should not be underestimated—allow many hundreds of hours!

What is certain, whatever the software and hardware approach that is chosen, is that the optimization must be directed to fulfil some pre-specified needs. There must be an optimization goal. Such a goal is commonly described in terms of an optimization criterion or function. A great number have been developed to be used in a wide variety of schemes and, as they are fundamental to most of the methods that follow, some of them will be examined next.

2.3. OPTIMIZATION CRITERIA

One of the most contentious aspects of HPLC optimization is that of designing suitable criteria for an optimization procedure. It is generally accepted that the assessment of a chromatogram is a multi-criterion process, as pointed out in Chapter 1, but just what aspects should be considered in attempting to decide whether one chromatogram is superior to another? For the purposes of optimization a chromatographic system can be represented[34] as shown in Figure 2.2. The sample is introduced to the chromatograph and is separated according to the controlled influences of parameters such as mobile phase composition and flow rates, column type and separation temperature. There will be other uncontrolled influences on the separation such as noise and drift

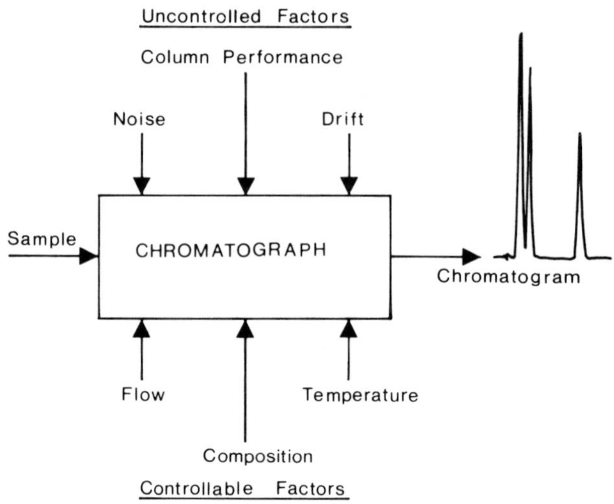

Figure 2.2. A schematic representation of the chromatographic system and the controlled (and hence changeable) and uncontrolled factors

which will affect the ultimate output signal—the chromatogram[35]. Additionally, although the column will have been chosen with some basic performance in mind, during the separation this performance may change (deteriorate) significantly in an unpredicted, and uncontrolled, manner. If all the parameters, and the consequent output, could be described exactly in a mathematical way, then no experimental optimization would be required. The chromatographic processes are so complex that, to date, such a mathematical model has proved elusive.[36,37]

What is known about the chromatographic system may be limited to a sparse set of semi-empirical rules which describe, in part, the effects of changing the controllable input parameters. Indeed, even less may be known if optimization is being carried out by a directed search routine (see Chapter 6). To realize fully automated optimization of HPLC separations, the controlling device (computer) must be in a position to judge the quality of any given separation. The computer can then either direct the chromatographer to the area of optimum chromatography or feed back information to the controlled input parameters such that progress is made towards the region(s) of optimum chromatography. To be able to do either of these things, some form of quality criterion, or function, is required.

The selection of a quality criterion will be dependent on the problem being tackled and the sample type being handled. In optimization terms it is possible to classify[38] samples as belonging to one of three types:

i. a mixture of a known number of essentially known compounds;
ii. a mixture of a known number of components with a number of (unknown)

background compounds; the analytically desired compounds are also available in a pure form without the background matrix;

iii. a mixture of an unknown number of unknown components.

Samples of the first two categories are relatively easily handled by many of the simpler quality criterion functions but, as will be seen, more complex separations, particularly those in the third category, require considerably more involved functions.

Since 1960 there have been numerous attempts to design quality criteria for chromatographic optimization. Giddings[39] describes a Total Overlap Method:

$$\phi = \Sigma \exp(-2R_i) \tag{2.5}$$

where R_i is the resolution between the ith peak pair.
Massart and Smits[40] used the informing power of a chromatogram consisting of n peaks:

$$P_{\text{inf}} = \sum_{i=1}^{n} \log_2 S_i \tag{2.6}$$

where S_i is the reciprocal of the fractional overlap between peaks i-1 and i. A (1/time) term may precede the main function to allow for the separation time.

Morgan and Deming[41] generalized Kaiser's[42] concept of peak separation (Figure 2.3) to produce the Chromatographic Response Function (CRF):

$$\text{CRF} = \sum_{i=1}^{n} \ln(P_i) \tag{2.7}$$

where P_i is the peak separation for n peak pairs. This relatively simple CRF has the merits of being easy to calculate if the peak separation is being determined manually. The peak separation turns out to be a difficult value to determine automatically. It will often be better in practice to make some assumptions about the symmetry of the peaks and to attempt to estimate resolution directly, particularly when smaller peaks are eluting on the tail of large peaks or when there is baseline drift.

Any of the three quality functions described so far may suffice for the optimization of samples from the first and, possibly, the second categories, but it was soon realized that all had restrictive limitations. A major problem is that

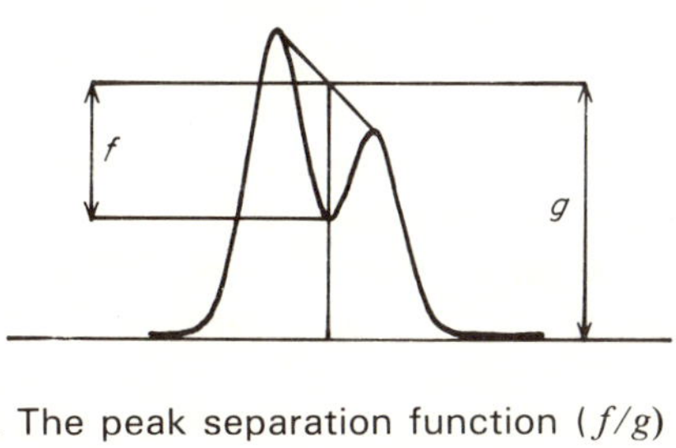

Figure 2.3. The peak separation function (f/g) of Kaiser[42]

separations can give equal values for the functions as peak order changes or as resolutions between different peak pairs changes. Additionally, it was recognized that no allowance had been made for time, an important consideration.

The time requirement was recognized by Watson and Carr,[43] who described the CRF as shown in equation 2.8:

$$\mathrm{CRF} = \sum_{i=1}^{n} \ln(P_i/P_0) + a(t_\mathrm{m} - t_\mathrm{n}) \tag{2.8}$$

where

n = the number of adjacent peak pairs;
P_0 = the desired peak separation;
P_i = the peak separation for the ith pair of peaks;
t_m = the maximum acceptable analysis time;
t_n = the elution time of the last peak;
a = an arbitrary weighting factor.

This function is constrained so that if $P_i > P_0$ then P_i is set equal to P_0. This constraint has the effect of neglecting peak pairs where the resolution achieved exceeds the desired resolution, allowing optimization to concentrate on peak pairs with inadequate resolution. A second constraint is to set t_m equal to t_n if t_m should exceed t_n: this will neglect the effect of time for all chromatograms that meet the time criterion.

The CRF in equation 2.8 signified a major improvement in quality criteria and it was further modified by Glajch *et al.*[44] Rather than describe a CRF, Glajch *et al.* chose to use the term chromatographic optimization function (COF):

$$\mathrm{COF} = \sum_{i=1}^{n} A_i \ln(R_i/R_{i\mathrm{d}}) + B(t_\mathrm{m} - t_\mathrm{n}) \tag{2.9}$$

where R_i is the resolution for the ith pair and $R_{i\mathrm{d}}$ is the desired resolution for the ith pair. The time terms t_m and t_n again refer to a desired maximum analysis time and the time of the last-detected peak, respectively. In addition to the weighting parameter B, a further arbitrary weighting, A_i, was added for each pair of peaks of interest to allow for individual peak pairs to be weighted differently according to the needs of the separation. An additional modification was to use resolution rather than peak separation on the argument that the performance of the peak separation function deteriorates rapidly for $P < 0.75$ ($R < 1.0$). This property of the peak separation function may be unduly restrictive where difficult separations are to be optimized. If all A_i values and all $R_{i\mathrm{d}}$ values are set the same for all peak pairs, then the COF equation 2.9 reduces essentially to the CRF equation 2.8. In use, the COF gives a single number to describe the chromatographic quality which tends towards zero as the optimum separation is reached; poor chromatograms will produce large, negative values for the COF.

These modifications to the quality function are a considerable improvement for many situations. Referring to the sample types described earlier, it is evident that the COF of Glajch *et al.* will certainly handle samples in the first two categories although it will be ineffective for the third category. Even for the first two categories, however, it still suffers from severe limitations. These limitations pertain to problems where peak crossover occurs, for which the COF will make no allowance, and identical values of the COF can result from chromatograms with different numbers of peaks.

Other workers also recognized the problems associated with the use of the peak separation number of Kaiser. Schoenmakers *et al.*[45] proposed using the product of all resolutions (R_s) between adjacent peak pairs in order to aim for a chromatogram in which the peaks were all evenly spaced:

$$R_s = \prod_{i=1}^{n} R_{s_{i+1,i}} \tag{2.10}$$

The lowest value of R_s has a dominant effect on the value of the resolution product, which at its highest value corresponds to an even spacing of all detected peaks throughout the chromatogram.

It was soon recognized that this criterion occasionally gave rise to inferior chromatograms and it was modified[5,46] to the form shown in equation 2.11, giving the relative resolution product (r):

$$r = \prod_{i=1}^{n-1} R_{s_{i+1,i}} \Big/ \left[\left(\sum_{i=1}^{n-1} R_{s_{i+1,i}}\right)/(n-1) \right]^{n-1} \tag{2.11}$$

where the denominator denotes the maximum possible value for ΠR_s within a given chromatogram. The value of the relative resolution product goes from zero for a coinciding peak pair to a maximum value of unity when all peak pairs are equally spaced throughout the chromatogram. Even with this relative resolution product there will be situations where the same result will be produced for quite different chromatographic performance, particularly as no time factor is involved. This latter point can be taken into account by including a further refinement in which an imaginary peak, indicated by $i = 0$, is considered to elute at a desired starting point in the chromatogram. The equation is now

$$r^* = \prod_{i=1}^{n-1} R_{s_{i+1,i}} \Big/ \left[\left(\sum_{i=0}^{n-1} R_{s_{i+1,i}}\right)/(n-1) \right]^{n-1} \tag{2.12}$$

In the denominator of this equation,

$$R_{s_{1,0}} = (t_1 - t_0)N^{1/2}/2(t_1 + t_0) \tag{2.13}$$

The criterion expressed by equation 2.12 thus aims at an even spacing of all peak pairs early in the chromatogram.

None of the functions considered so far has been ideally suited to the optimization of samples in the third category. For samples of this type, a quality function is needed that can be biased towards the discovery of more peaks. In other words, given two chromatograms for the same sample, the chromatogram containing the more peaks must be evaluated as being better. This should be the case even if the separation quality between a given set of peaks appears worse. With this in mind, Vajda and Leisztner[38] proposed a new function with the form

$$\mathrm{CEF} = \sum_{i=1}^{n} A_i + \frac{\sum_{j=1}^{n-1} B_j P_j}{n-1} \tag{2.14}$$

where

A_i = 1 for the ith peak if it is analytically significant, otherwise $A_i = 0$;
B_j = 1 for the jth peak pair if at least one of the peaks of the pair is analytically significant, otherwise else $B_j = 0$;
P_j = the peak separation for the jth peak pair;
n = the number of peaks in the chromatogram.

The CEF function in equation 2.14 will produce its largest value for the 'best' chromatogram; this will be equal to $n+1$ in the case of baseline separation between every peak. The advantages of this function result from its evaluation of a chromatogram with one more peak as being better than any other chromatogram. The CEF can be used to evaluate samples of the third category: parameters A_i and B_j will be set to unity.

As described, the CEF does not aim at a minimum separation, but such a requirement can be incorporated into it. It will also be seen that there is no reference to time. The originators of the function argued that time could not be built in easily but there seems no obvious reason why a time consideration could not be incorporated into the function with a suitable weighting.

Wegscheider *et al.*[47] designed a CRF that did include a time penalty term, in addition to a term to take into account noise on the chromatogram. Their function (equation 2.15) is a simple extension of equation 2.7:

$$\mathrm{CRF} = 1/t_{95} \prod_{i=1}^{n-1} f_i/(g_i + 2n_i) \tag{2.15}$$

where f_i and g_i are as described by Kaiser for the determination of the peak separation factor (Figure 2.3). The term $2n_i$ is added to take account of imprecision in the determination of separation factor because of noise and t_{95} (i.e. 95% of the analysis time) avoids excessively long retention times. There is, however, no automatic accounting of peaks.

It was the desire to design an optimization function that would allow resolution, time and the number of peaks detected to be considered simultaneously that led Berridge[33] to develop the CRF shown in equation 2.16:

$$\mathrm{CRF} = \sum_{i=1}^{n} R_i + n^a - \mathrm{b}|T_\mathrm{A} - T_\mathrm{n}| - c(T_0 - T_1) \tag{2.16}$$

where

R_i = the resolution between adjacent peak pairs;

n = the number of peaks detected;
T_I = the retention time of the first detected peak;
T_0 = a specified minimum desired retention time for the first detected peak;
T_A = the maximum desired retention time for the last eluted peak;
T_n = the retention time of the last detected peak;
a, b, c = operator selectable weightings which are usually set in the range 0–3.

It has been argued[48] all these functions suffer from limitations which can be relatively easily overcome. For example, functions which incorporate a logarithmic term approach zero at their maximum but no indication is given whether a large number of peaks have been well separated or whether just a few peaks are poorly separated. Additionally, peak crossovers cause a major problem since the functions will be zero, causing the location of false optima. Many functions do not consider the number of peaks detected during the optimization, or the number must be known beforehand. The new function of Van Hare and Rogers[48] was designed to overcome these limitations by incorporating both a peak counting part and a peak separation part. The function has two forms:

$$\mathrm{SEP}_P = n + \sum_{i=1}^{n} P_i + 0.5\, N_s \tag{2.17a}$$

$$\mathrm{SEP}_V = n + \sum_{i=1}^{n} (1 - V_i) + 0.5\, N_s \tag{2.17b}$$

where
n = the number of peaks detected;
P_i = Kaiser's peak separation function;
N_s = the number of shoulders as determined by a second derivative method;
V_i = the valley-to-peak ratio.

Equations 2.17a and 2.17b are reported to give very similar results and to be applicable to all chromatographic methods, requiring no prior information. One of the virtues is said to be that the function can be expanded to include weighting terms along the lines of those incorporated in equation 2.16—an almost certain necessity for many HPLC applications. As given, the function suffers greatly from the need to determine the second derivative of the chromatogram with respect to time. This will require that all the digitized data are stored for post-chromatographic processing by suitable smoothing and differentiating algorithms.

The optimization functions discussed above are only a selection of those that are available in the literature; a more detailed list is presented in Table 2.1, together with the uses to which they have been put.

TABLE 2.1. Chromatographic quality evaluation functions

Quality function	Experimental variables	Optimization method	Ref.
(*a*) *Informing power and separation number:*			
$P_{\mathrm{inf}} = \sum_i^n \log_2 S_i$			40
$P_{\mathrm{inf}} = \frac{1}{t} \sum_i^n \log_2 S_i$			
$S_N = \sum_i^n \log_2 P_i$			49
(*b*) *Chromatography response and optimization functions:*			
$\mathrm{CRF} = \sum_i^n P_i$	GLC	Simplex search	41
$\mathrm{CRF} = \frac{1}{t} \sum_i^n P_i$			
$\mathrm{CRF} = \sum_i^n a_i \ln P_i$			
$\mathrm{COF} = \sum_i^n a_i \ln (R_i/R_{\mathrm{d}}) + b(t_{\mathrm{m}} - t_{\mathrm{n}})$	Composition of quaternary mobile phase	Simplex design, empirical model and numerical solution	44
$\mathrm{CRF} = \sum_i^n R_i + n^a - b\lvert t_{\mathrm{m}} - t_{\mathrm{n}}\rvert - c(t_{\mathrm{o}} - t_1)$	Composition of ternary mobile phase, temperature, flow rate, pH	Simplex search	33 50
$\mathrm{CRF} = \sum_i^n \ln(P_i/P_{\mathrm{o}}) + a(t_{\mathrm{m}} - t_{\mathrm{n}})$	Gradient parameters and flow rate	Simplex search	43

Continued

TABLE 2.1. *(Continued)*

$\mathrm{CRF} = \lvert t_m - t_m^* \rvert / t_m + \lvert R_s - R_s^* \rvert / R_s$	Composition of binary mobile phase	Numerical solution/computer prediction	51
$\mathrm{CRF} = \frac{1}{t} \prod_i f_i/(g_i + 2n_i)$	Concentration of organic modifier and buffer, pH	Simplex search	52
$\mathrm{CRF} = \sum_i^n \ln(f_i/g_i) - 100\ (M - n)$	Concentration of organic modifier, pH	Simplex search	53
$F_{obj} = \sum_i^n [10(1.5 - R_i)]^2$	Composition of ternary mobile phase	Simplex search	54
$F_{obj} = \sum_i^n 100e^{(1.5 - R_i)} + (t_m - t_n)^3$	Composition of ternary mobile phase	Simplex search	55
$\mathrm{SEP}_p = n + \sum_i^n P_i + 0.5N_s$	GLC		48
(*c*) *Resolution product:*			
$R_s = \prod_i^n R_i$ $r^* = \dfrac{\prod_{i=1}^{n-1} R_s}{\left[\sum_{i=0}^{n-1} R_s/(n-1)\right]^{n-1}}$	Composition of ternary and quaternary mobile phase	Iterative mixture design	5 46
(*d*) *Separation factor:*			
$S = \sum_{a<b} (t_b - t_a)/(t_a + t_b)$	pH	Window diagram	56

2.4. SUMMARY

The choice of hardware and software is fundamental to the success of automated separation optimization. There is no single chromatographic system that is better than any other for the purposes of automated optimization. Questions such as the reliability and ease of servicing will play just as important a part in the purchase assessment as cost and capabilities. What is required is a chromatograph that can be configured to mimic that shown in Figure 2.1, namely:

2–4 solvent capability under remote control;
auto-injector;
robust separation column;
detector matched to the needs of the separation;
integrator or data handling system;
system controller, ideally a computer, with good communication.

When it comes to considering criteria, given such an enormous number of optimization functions with such a diverse range, there is the inevitable problem of having to decide which one to select and apply. Some help may come from the uses to which an individual function has been put (Table 2.1) and the use for which it is desired. There have also been a number of investigations into the relative performance and merits of particular functions.[35,57,58] Debets *et al.*[35] concluded that none of the published methods available in 1983 is unconditionally useful in directing the automated optimization of separations. The problems arise because, it is argued, for any given value of a function there may be a large, if not infinite, number of separations that can produce this value. Faced with a separation problem and needing a function, it is suggested that a function should be selected which is (a) easy to calculate and (b) appropriate to the separation under consideration. Thus the optimization functions described in equations 2.7, 2.9, 2.12 and 2.16 are highly commended.

2.5. REFERENCES

1. H. M. McNair, *J. Chromatogr. Sci.*, **22** (1984) 521.
2. T. Greibrokk, A. L. Blilie, E. J. Johansen and E. Lundanes, *Anal. Chem.*, **56** (1984) 2681.
3. D. D. Blevins, M. F. Burke and V. J. Hruby, *J. Chromatogr. Sci.*, **21** (1983) 490.
4. P. Jandera, J. Churacek and H. Colin, *J. Chromatogr.*, **214** (1981) 35.
5. A. C. J. H. Drouen, *Ph D Thesis*, Technische Hogeschool Delft, The Netherlands, 1985.
6. J. G. Stewart and P. A. Williams, *J. Chromatogr.*, **198** (1980) 489.
7. A. P. Goldberg, *Anal. Chem.*, **54** (1982) 342.
8. I. S. Lurie and S. M. Carr, *J. Liq. Chromatogr.*, **6** (1983) 1617.
9. A. M. Krstulovic and P. R. Brown, *Reversed-Phase High-Performance Liquid Chromatography, Theory, Practice and Biomedical Applications*, Wiley, New York, 1982.

10. J. F. K. Huber, M. Pawlowska and P. Markl, *Chromatographia*, **19** (1984) 19.
11. S. H. Hansen, P. Helboe and M. Thomsen, *J. Pharm. Biomed. Anal.*, **2** (1984) 167.
12. J. R. Benson and D. J. Woo, *J. Chromatogr. Sci.*, **22** (1984) 386.
13. W. H. Pirkle, M. H. Hyun and B. Bank, *J. Chromatogr.*, **316** (1984) 584.
14. W. H. Pirkle, M. H. Hyun, A. Tsipouras, B. C. Hamper and B. Bank, *J. Pharm. Biomed. Anal.*, **2** (1984) 173.
15. B. A. Bidlingmeyer and F. V. Warren, *Anal Chem.*, **56** (1984) 1583A.
16. H. J. Issaq, *J. Liq. Chromatogr.*, **6** (1983) 1375.
17. H. J. Issaq, *J. Liq. Chromatogr.*, **7** (1984) 883.
18. E. Katz, K. L. Ogan and R. P. W. Scott, *J. Chromatogr.*, **289** (1984) 65.
19. J. L. DiCesare, *Ind. Res. Dev.*, **25**(4) (1983) 110.
20. F. Erni, *J. Chromatogr.*, **282** (1983) 371.
21. K. Jinno, J. B. Phillips and D. P. Carney, *Anal. Chem.*, **57** (1985) 574.
22. 'Focus', *Anal. Chem.*, **54** (1982) 327A.
23. P. M. Lyne and K. F. Scott, *J. Chromatogr. Sci.*, **19** (1981) 547.
24. P. M. Lyne and K. F. Scott, *J. Chromatogr. Sci.*, **19** (1981) 599.
25. P. Tarroux and T. Rabilloud, *J. Chromatogr.*, **248** (1982) 249.
26. G. I. Ouchi, *Res. Dev.*, **26**(2) (1984) 150.
27. M. Bogdan and P. K. Warme, *Comput. Appl. Lab.*, **1** (1984) 21.
28. J. M. Anderson, *J. Liq. Chromatogr.*, **6** (1983) 2809.
29. G. W. Inman, K. D. Holmes, G. W. Martin, T. H. Barret and S. L. Varghese, *Int. Lab.*, Nov/Dec (1983) 10.
30. D. L. Gustine and J. McCulloch, *J. Chromatogr.*, **316** (1984) 407.
31. J. P. Bounine, G. Guiochon and H. Colin, *J. Chromatogr.*, **298** (1984) 1.
32. J. J. Kirkland, W. W. Yau, H. J. Stoklosa and C. H. Dilks, *J. Chromatogr. Sci.*, **15** (1977) 303.
33. J. C. Berridge, *J. Chromatogr.*, **244** (1982) 1.
34. D. L. Massart, A. Dijkstra and L. Kaufman, *Evaluation and Optimisation of Laboratory Methods and Analytical Procedures*, Elsevier, Amsterdam, 1978.
35. H. J. G. Debets, J. J. Weyland and D. A. Doornbos, *Anal. Chim. Acta*, **150** (1983) 259.
36. J. C. Smit, H. C. Smit and E. M. De Jager, *Anal. Chim. Acta*, **122** (1980) 1.
37. J. C. Smit, H. C. Smit and E. M. De Jager, *Anal. Chim. Acta*, **122** (1980) 151.
38. J. Vajda and L. Leisztner, in H. Kalasz (Ed.), *New Approaches in Liquid Chromatography*, Analytical Symposium Series, Vol. 16, Elsevier, Amsterdam, 1983, p. 103.
39. J. C. Giddings, *Anal. Chem.*, **32** (1960) 1707.
40. D. L. Massart and R. Smits, *Anal. Chem.*, **46** (1974) 283.
41. S. L. Morgan and S. N. Deming, *J. Chromatogr.*, **112** (1975) 267.
42. R. E. Kaiser, *Gas Chromatographie*, Geest and Portig, Leipzig, 1960.
43. M. W. Watson and P. W. Carr, *Anal, Chem.*, **51** (1979) 1835.
44. J. L. Glajch, J. J. Kirkland, K. M. Squire and J. M. Minor, *J. Chromatogr.*, **199** (1980) 57.
45. P. J. Schoenmakers, A. C. J. H. Drouen, H. A. H. Billiet and L. de Galan, *Chromatographia*, **15** (1982) 688.
46. A. C. J. H. Drouen, H. A. H. Billiet, P. J. Schoenmakers and L. de Galan, *Chromatographia*, **16** (1982) 48.
47. W. Wegscheider, E. P. Lankmayr and M. Otto, *Anal. Chim. Acta*, **150** (1983) 87.
48. D. R. Van Hare and L. B. Rogers, *Anal. Chem.*, **57** (1985) 628.
49. W. A. Spencer and L. B. Rogers, *Anal. Chem.*, **52** (1980) 950.
50. J. C. Berridge, *Analyst (London)*, **109** (1984) 291.
51. K. Jinno and K. Kawasaki, *J. Chromatogr.*, **298** (1984) 326.
52. W. Wegscheider, E. P. Lankmayr and K. W. Budna, *Chromatographia*, **15** (1982) 498.

53. J. H. Nickel and S. N. Deming, *Liq. Chromatogr. Mag.*, **1** (1983) 414.
54. A. S. Kester and R. E. Thompson, *J. Chromatogr.*, **310** (1984) 372.
55. D. L. Dunn and R. E. Thompson, *J. Chromatogr.*, **264** (1983) 264.
56. P. Jones and C. A. Wellington, *J. Chromatogr.*, **213** (1981) 357.
57. H. J. G. Debets, B. L. Bajema and D. A. Doornbos, *Anal. Chim. Acta*, **151** (1983) 131.
58. J. W. Weyland, C. H. P. Bruins, H. J. G. Debets, B. L. Bajema and D. A. Doornbos, *Anal. Chim. Acta*, **153** (1983) 93.

3 Simple Methods and Pragmatic Approaches

3.1. INTRODUCTION

This chapter gathers together a number of methods which can provide a quick and easy solution to the optimization of a separation. The methods might also form a starting point for a more detailed optimization procedure, using one of the approaches which will be discussed in subsequent chapters. Many of these simpler techniques can be carried out with inexpensive hardware and many require only relatively simple computation.

The possibilities of using a standardized strategy will be examined first. This strategy aims to minimize the number of columns and mobile phases that need to be investigated. The idea works particularly well for basic compounds, for example drugs, and has recently been extended to neutral and acidic molecules. Faced with separating many different mixtures of compounds, more 'universal' methods which are also suitable for automation are needed. Detection is often a problem here but, fortunately, most organic compounds absorb UV light of short wavelengths (190–210 nm). If solvents and reagents can be obtained sufficiently pure, these low wavelengths can be used. Separation of mixtures with a wide range of polarities can then be investigated using automated sequences of isocratic runs, or gradients.

Gradient separations provide the starting point for many automated optimization schemes. From a suitably designed gradient separation it is possible to estimate directly conditions for an isocratic separation. There are different suggestions as to how this can be achieved in practice and a number of these will be considered. The use of gradient elution separations for the prediction of isocratic conditions leads into the question of how gradient separations themselves may be optimized. There are some fairly simple 'rules of thumb' which can be tried; more exhaustive methods will be considered in chapters devoted to specific schemes.

Finally, some more pragmatic approaches, including methods which aim to predict the separation on the basis of theory alone, will be considered. There is a wide range of approaches which aim to integrate theoretical principles with the minimum of experimentation. Some are simple and readily implemented. Others are complex and actually require considerable experimentation to provide the basis for comprehensive optimization.

3.2. STANDARDIZED METHODS

3.2.1. A Single-column Standardized Strategy

The first problem encountered when developing a new chromatographic method is that of just where one should start. There is such an enormous range of stationary phases and mobile phases to be investigated. What is the 'best' column to use? What should the first experiment be? (Some help in answering these questions is available in Chapter 2.) What is required, perhaps, is a restricted set of systems (i.e. stationary phase plus mobile phase). This was the approach taken by Massart and co-workers.[1,2] They began by examining 16 chromatographic systems for the separation of 100 basic drugs. The assessment of preferred HPLC systems was made on the basis of their relative discriminating power. What emerged was that the majority of the compounds could be separated from each other using just two systems. The two systems have the major added advantage that they both use the same column stationary phase, namely a cyano-bonded phase. Hence for the one column only, two mobile phases are needed: (a) n-hexane (or n-heptane)/methylene chloride/acetonitrile/propylamine (50:50:25:0.1); (b) acetonitrile/water/propylamine (90:10:0.01). Naturally, it is not suggested that these two mobile phases will succeed for all mixtures; what they may do is provide a good starting point for further optimization. The advantages of adopting this method are obvious. The need to choose a column is eliminated, since all separations require just a single column, and the final optimization of separation conditions can be restricted simply to the fine tuning of the solvent ratios. Such a strategy is ideally suited to automated optimization using relatively modest equipment since most controllers are able to handle at least the two basic compositions. For fully automated applications, a four-solvent system can handle both modes of chromatography if the amine modifier is pre-blended with the other solvents. There will be some changes in amine content as the solvent compositions change, but this should not be too detrimental to the optimization result. Provided that due care is taken to avoid the problems of generating immiscible solvent mixes, it should be possible to investigate both reversed- and normal-phase separations in a single experimental run.

This standard approach becomes even more attractive when combined with a complementary standardized extraction method.[3,4] With such a process, the analyst can tackle a wide variety of problems with a high degree of confidence that first experiments will provide good data, if not the final answer. The value of this approach can be seen in the literature applications, which are many.[1-7] For example, drugs in formulations have been tested for conformity with their label claims, for stability and for the determination of degradation products. There are also examples of applications to the determination of drugs in biological fluids such as plasma, urine and saliva. The method has also been applied successfully to cosmetics formulations.

Application to neutral and acidic compounds is also possible, still using just a cyano-bonded stationary phase.[8] To achieve this, however, for both normal- and reversed-phase chromatography requires an extension of the number of solvents to six rather than four. The six solvents are water and n-hexane, for adjusting the solvent strength in reversed- and normal-phase separations, respectively, with methanol, acetonitrile, tetrahydrofuran and dichloromethane as selectivity adjusters.

Before embarking on the simplified strategy, it is necessary to decide on the chromatographic mode which will be used. In general, polar substances should be first examined by reversed-phase and non-polar substances by normal-phase chromatography. There will be many exceptions to this guide and it will frequently be advantageous to examine mixtures by both separation modes. The acid–base nature of the compounds should also be considered. For acidic compounds, 1% acetic acid should be added to all solvents. For basic compounds the addition of propylamine is recommended. While 0.1% is suggested in normal-phase separations, this concentration can lead to excessive dissolution of the column packing in reversed-phase separations and 0.01% is a better concentration.

The solvent strength required to elute the compounds of interest is determined following a gradient separation. For normal-phase chromatography the gradient would be from n-hexane to dichloromethane whereas in reversed-phase chromatography a gradient from water to methanol would be employed. The precise method of estimating the required solvent strength will be described in detail in Section 3.3. This first isocratic composition will use methanol and water (reversed phase) or n-hexane and dichloromethane (normal phase). Two other binary, isocratic compositions are then estimated, using mixtures of the two remaining selectivity adjusting solvents with water or n-hexane as appropriate. It is likely that the acetonitrile/n-hexane mobile phase will not be miscible and a separation with this mobile phase is not carried out. Finally, three ternary mixes with same solvent strength are prepared from 1:1 mixtures of the binary solvents. This provides six mobile phases for reversed-phase and five for normal-phase separations. The experimental design is referred to as a (partial) simplex lattice design and this formal experimental design will be considered in greater detail in Chapter 4, where it forms the basis for a comprehensive approach to separation optimization.

The only examples and applications of the extended, simplified strategy are in the original paper. Benzamides, local anaesthetics, phenothiazines, tricyclic antidepressants, benzodiazepines, corticosteroids, sulphonamides, barbiturates and food preservatives were examined and found to be amenable to this simplified separation strategy.

3.2.2. Universal Methods

The twin aims of the so-called 'universal' methods are to reduce, or possibly eliminate, manual method development and to provide chromatographic systems that can be applied to wide varieties of molecules with different polarities and detectabilities. The first aspect is achieved simply by the application of automation. The second is slightly more difficult but advantage can be taken of the fact that most organic molecules absorb low-wavelength UV light.

Some of the very earliest attempts[9,10] at automated separation optimization were, in fact, very close to approaching universal applicability. Just because these methods are old does not mean that they are outdated; they actually have much to commend them. Given a microcomputer-controlled chromatograph, much can be done to achieve automated optimization with a minimum of programming. For example, even a modest two-solvent system can be set up to chromatograph automatically a mixture with a number of different solvent compositions. At the end of the run the chromatographer can select the best (although not necessarily the optimum) separation conditions by inspection. This was the philosophy behind the earliest approaches to automated separation development, being given the name 'programmed runs'.[9,10] The approach can be extremely successful where detection conditions are known, and is highly recommended. If a programmable chromatograph is available, what is there to lose by such an approach for a new mixture, especially if the run is carried out unattended, overnight? Where detection conditions are unknown, by using UV detection at as low a wavelength as possible (e.g. 190 nm), the method becomes essentially universal. One should not forget the possibilities of increasing detection universality that are offered by diode-array detectors. With most diode-array detectors it is possible to add the signals from all the individual diodes to produce an output representing the sum of all wavelengths over the UV, and possibly visible, spectrum.

The combination of low-wavelength detection and a series of isocratic separations has been termed 'sequential isocratic step' (SIS) chromatography[11] and is well suited to reversed-phase separations. The SIS method involves the use of a computer-controlled chromatograph to carry out sequentially a series of isocratic separations using acetonitrile and an aqueous buffer. The steps are placed at intervals of 20% acetonitrile, starting at 100% acetonitrile, each covering a range of about ten k' units. In just six runs, the whole polarity range, from 100% acetonitrile to 100% water, can be covered. Near universal detection can be achieved by using UV light at 190 nm, although this does place some limitations on both the solvent/reagent qualities and the performance of the detector. It is not essential to use a wavelength of 190 nm and, if knowledge is to hand about the solutes, other wavelengths could be used with the advantages

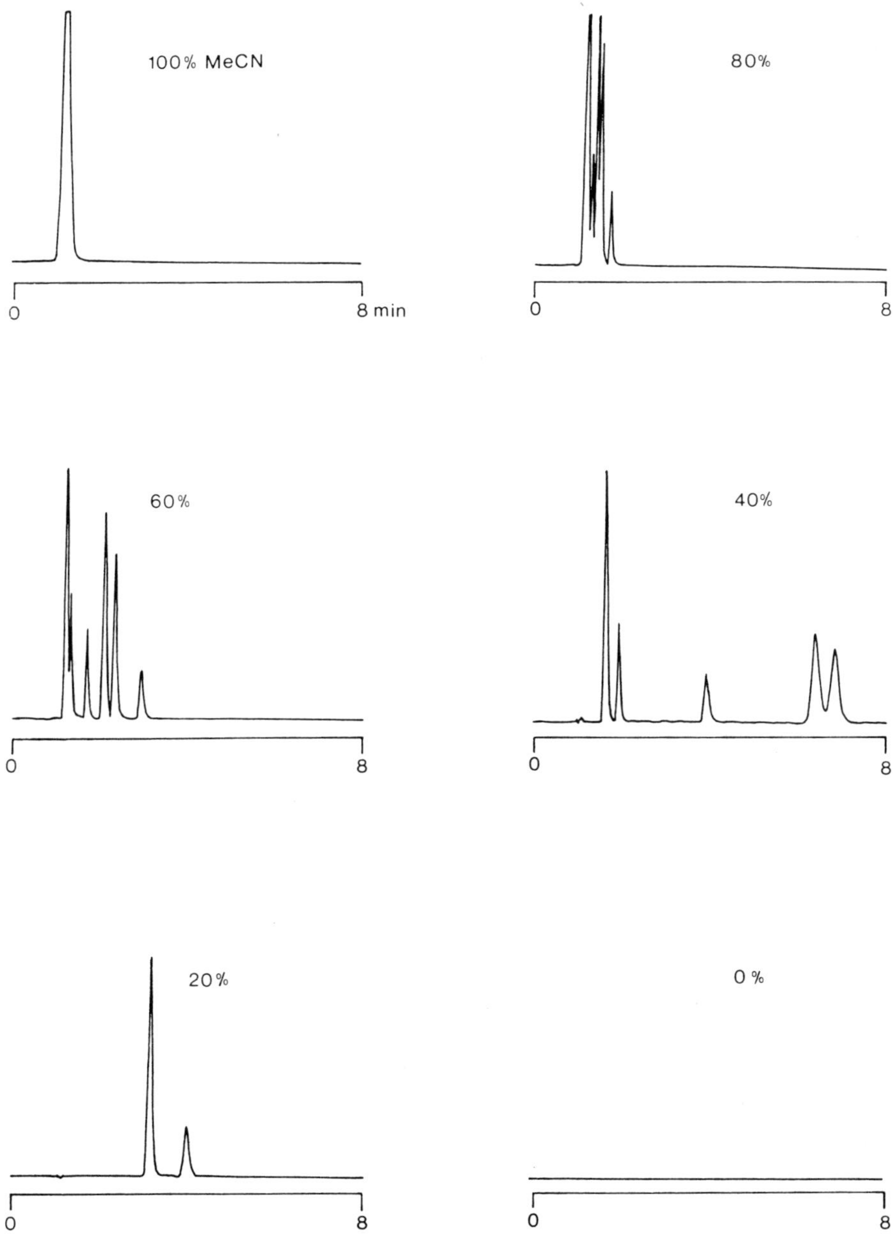

Figure 3.1. Sequential isocratic step (SIS)[11] of the six-component test mixture using 20% decrements of acetonitrile

of greater selectivity. Additionally, if higher wavelengths can be chosen, other solvents such as methanol can be used as the modifier. Figure 3.1 shows the SIS approach applied to the six-component test mix that is being used to illustrate some of the methods being described. In this case acetonitrile was used as the organic modifier. Detection was effected with a diode-array detector operating with a wide bandwidth to cope with solutes having very different spectral characteristics.

Perhaps no single separation shown in Figure 3.1 would be regarded as optimum but, in less than one hour, a good indication of the required composition was achieved. Figure 3.1 also suggests a warning about the use of SIS methods. In half of the separations, at least one of the peaks was not eluted before the chromatogram was completed. Hence there is a danger that late-eluting peaks may appear in a subsequent chromatogram and, although their presence will probably be given away by their extended peak widths, there is a danger of misinterpretation of the chromatogram.

SIS methods have been applied successfully to the separation of phthalate esters, for the comparison of batches of production solvents and for ion-pair separations.[11] Software for a Hewlett-Packard 1084B is provided in ref. 11, but the setting up of equivalent methods for other chromatographs should be easily accomplished.

One of the reasons for using SIS HPLC is to avoid the problems that can arise from gradient elution with low-wavelength detection. These problems include baseline drift and artifact, or ghost, peaks during the run. The ghost peaks usually originate from the elution of impurities collected on the column from the water, or the organic phase, being eluted later in the run.[12] If the mobile phase components can be made sufficiently pure, then universal gradient methods become available. There will always be some gradient baseline displacement owing to the inherent differences between the UV characteristics of the two solvents. A means of reducing ghost peaks for acetonitrile/water gradients is to clean the acetonitrile by using an in-line column of acidic alumina (Brockman activity 1) to remove amine-like substances and to clean the water by using an in-line column of Chromosorb 101 (styrene–divinylbenzene porous polymer).[13] The water column should be cleaned daily by flushing with acetonitrile. By careful cleaning of the solvents, gradients from 100% water to 100% acetonitrile can be run at 210 nm and at high sensitivity. If a buffer is not being used in the aqueous phase it is possible to clean the water very successfully simply by passing it through a polytetrafluoroethylene (Teflon) tube around a 189 nm UV irradiator.[14] If gradient drift as a result of UV mismatch is still a problem, this can be minimized by blending in proportions of nitrous oxide and helium[15] to match the UV absorbances of the two solvents. Where methanol is required as the modifier, formamide or sodium nitrate[16] can be used to compensate for absorbance mismatches. For pH control, buffers of 7.5 mM triethylamine phosphate at pH 3 or 2.5 mM ammonium borate at pH 9 can be used, although reduced column lifetime must be expected if a pH as high as 9 is used for extended periods. If it is desired to use wavelengths below

210 nm, then this is possible with care.[17] Triethylamine can still be used as the buffering agent and the gradient absorbance mismatch is now compensated for by adding a mixture of nitrous oxide and carbon dioxide. By varying the concentration of triethylamine it is possible to control the pH of the mobile phase within precise limits; the saturation of the mobile phase with carbon dioxide will effectively control the final pH. The experimental difficulties which arise with this method are associated with the problems of pumping a solvent saturated with gases. It is usually recommended that mobile phases be degassed prior to use. Where low concentrations of gases are being used to compensate the UV baseline, then degassing is not a major problem. With the carbon dioxide/nitrous oxide system due precautions to reduce bubble formation, and to trap any that may be formed, have to be taken. If only restricted gradients turn out to be necessary, e.g. 5–25% acetonitrile, wavelengths as low as 185 nm can be used,[18] although at these wavelengths it will not be possible to use a buffer and the optical system will need purging of oxygen.

Three solvent systems (pH 3 buffer, unbuffered and pH 9 buffer) plus acetonitrile are therefore available for near universal gradient separations at low wavelengths with high sensitivity. These gradients can be used for the 'first-time' resolution of components that are amenable to such separations. Of course, not all are. For example, it may be necessary to increase the selectivity of a separation by employing an ion-pair reagent. The obvious way to do this would be to mix the reagent with one, or both, of the solvents. However, in attempting to devise universal, automated methods for all sample types, it is useful to be able to investigate both cation- and anion-paired separations, as well as non-paired separations.

A proposed means of managing the automated investigation of ion-pair separations is to load a concentrated solution of an ion-pairing reagent into simple buffered eluents[15,19] using the capabilities that are available with many modern autosamplers. This technique is known as 'injection loading'. It is a complicated procedure to explain (and perhaps to optimize in practice), but it is of interest because of its flexibility. In practical terms the sample is still injected under conventional chromatographic conditions and elution is commenced in the normal way. However, just before the sample injection, an injection of a concentrated solution of the reagent under investigation is made. A zone of the reagent passes down the column with the mobile phase and it will interact with the solutes. Within this microenvironment some solutes will be affected whereas others will not. Inspection of chromatograms will permit the identification of the interacting solutes. Additional benefits that may accrue are that by local modification of the mobile phase it may be possible to modify just a small part of the total separation. This would leave most of the components unaffected, with the result that otherwise unobtainable selectivity enhancements can be achieved. Where ion-pair separations are being considered, a useful sequence of runs might be reversed-phase followed by cation-pairing and finally anion-pairing. With this sequence much information is available about the nature and behaviour of the solutes, particularly if components can be identified for all chromatograms.

Software for ion-pair injection loading is described in ref. 15 for a Hewlett-Packard 1084B HPLC system and again it is a relatively simple task to write a sequence for any other suitable instrument. Where an advanced autosampler is not available, injection loading can be attempted by injecting a concentrated solution of the reagent first with as short a run time as is possible. The mixture is injected immediately afterwards with the absolute minimum of delay. When carried out with a fully automated system, the injected loading method is likely to be highly reproducible and suitable for quantitative work.

Injection loading is not restricted to ion-pair loading and can be used to modify both isocratic and gradient separations.[19] As an example, it is not necessary to make the injection at the start of the run. During a gradient run an injection of water could be made to slow the gradient at a particular point. Rather than describe this use of the idea as 'injection loading', it has been described as 'solvent pulsing'. In these situations the method can become complex and the results can be difficult to predict. One should be aware also that injection loading and/or solvent pulsing suffer from a distinct lack of portability; it may be very difficult to transfer the method on to a chromatograph of completely different design as the timings may be substantially different.

3.3. ISOCRATIC SEPARATIONS BASED ON AN INITIAL GRADIENT SEPARATION

Gradient elution can be used as a very powerful 'scouting' technique for unknown samples. Indeed, a number of automated optimization methods rely on the results of an initial gradient separation to establish suitable boundary conditions for further isocratic optimization (see Chapters 4 and 6). A gradient separation provides an immediate picture of the sample nature and, from that separation, the necessary solvent strength for an isocratic separation can be calculated. The calculations may lead to the conclusion that an isocratic separation cannot be readily achieved within a reasonable time scale, which can save a lot of wasted time and frustration. Gradient separations are most frequently applied to reversed-phase chromatography and thus most methods are restricted to this mode.

If gradient separations are to be used as an approach to the prediction of isocratic separations, it is important to bear in mind that there are a number of experimental factors and errors to be taken into account.[20,21] Any errors or malfunctioning of the solvent delivery system will obviously lead to irreproducibility of gradient compositions. There are still more fundamental problems with gradients which must be considered. The composition that is delivered to the head of the column at any given time is dependent on whether low-pressure (i.e. one pump plus solvent proportioning valves) or high-pressure (i.e. two or more high-pressure pumps, each delivering a constant composition) gradient formation is being used. Solvent compressibilities, heats of mixing and volume changes following mixing will all play their part in frustrating attempts to deliver the nominal gradient to the column. Another factor to be considered is the inherent volume delays between the component which generates the gradient and the separation column. Many systems using high-pressure gradient generation include a stirred mixing chamber after the pumps to ensure that the mobile phase is homogeneous. The volumes of such chambers vary enormously but the effective volume, including the associated tubing, can exceed 2 ml. At a flow rate of 1 ml/min, any requested change in mobile phase composition will take at least two minutes to reach the column! However, in using gradients as a quick and easy scouting method, which will perhaps be followed by a more detailed experimental design, such errors can be accounted for and thus tolerated.

The first method of transferring gradient separations to isocratic separations comes from the studies of Snyder and co-workers.[22,23] If the solvent strength is changed in direct proportion to time this produces a linear gradient profile—a so-called linear solvent strength (LSS) gradient. For LSS gradients the inlet k' (k_i) values for individual solutes will decrease during the gradient elution program according to

$$\log k_i = \log k_0 - b(t/t_0) \tag{3.1}$$

where k_0 is the k' value expected for the solute using isocratic elution in the solvent composition corresponding to the start of the gradient elution program and t_0 is the column dead time. The term 'inlet k' values' refers to the isocratic k' for the solutes in a mobile phase with the given inlet composition at time t after the gradient has commenced. The parameter b is a measure of the slope of the gradient and remains constant during an LSS separation. It can be shown[22] that the retention time, t_g, of a solute during the gradient can be found from

$$t_g = (t_0/b)\log(2.3k_0b + 1) + t_0 + t_d \tag{3.2}$$

where t_d is the delay time of the system, corresponding to the time taken for a requested change in composition actually to reach the top of the column. The instantaneous k' value of a solute as it leaves the column (k_f), that is, the k' which would be expected in the solvent composition accompanying the peak, can be estimated from

$$k_f = 1/(2.3b + 1/k_0) \tag{3.3}$$

In isocratic separations the capacity factor is a function of the (constant) solvent strength of the mobile phase. If a binary mobile phase only is considered, the solvent strength is dependent on the volume fraction of a modifier (ϕ_b) and its solvent strength parameter, S. For mobile phases of water with methanol or acetonitrile and for a large number of solutes, equation 3.4 has been found to be valid:

$$\log k' = \log k_w - S\phi_b \quad (3.4)$$

where k_w is the isocratic k' value in pure water. This equation is not strictly obeyed in all reversed-phase systems but the consequences of any practical deviations are small in the current context. Since LSS gradients have been defined as gradients with constant changes in solvent composition:

$$\phi_b = \phi_0 + \phi' t \quad (3.5)$$

where ϕ_0 is the starting value of ϕ_b and ϕ' is the rate of change of ϕ_b with time.

The best value of ϕ' is, of course, dependent on the particular mixture so, for an unknown sample, some approximations are necessary. What has been predicted, and found experimentally,[22,23] is that a value of approximately 0.2 for b is optimum in generating a gradient profile that will provide good resolution and sensitivity. From equations (3.1), (3.4) and (3.5) it can be shown that

$$\phi' = b/(St_0) \quad (3.6)$$

All that remains to be established is a suitable value of S and, for an initial gradient scouting with methanol/water mixes, a value of 3 will suffice. In this particular instance, the desire is not to optimize the gradient, but to establish an approximate solvent strength for chromatographing the mixture under consideration. The value of 3 is adequate for small molecules but it has been pointed out that S can be related to the molecular weight of the solute[24] and a higher value (4–8) may be more appropriate for higher molecular weight molecules (i.e. greater than 500 daltons).

In practical terms, the following scheme is recommended for using an initial gradient separation to indicate isocratic separation composition:

(i) Run an initial linear gradient from 100% water to 100% methanol. The rate of change of methanol concentration (the gradient slope) is calculated from equation 3.6 by setting $b = 0.2$ and $S = 3$, i.e.

$$\%\ \mathrm{MeOH/min} = 100 \times 0.2/(3 \times t_0)$$

with t_0 expressed in minutes. For example, a column having a dead time of one minute will require a 6.7% increase in methanol concentration per minute.

(ii) Calculate the k_f value (the instantaneous k' value of the peak in the exit composition) of the band of interest using equation 3.3. If the elution time of the band is fairly long, k_0, the isocratic capacity factor in the initial solvent composition (water in this example), will be large and equation 3.3 reduces to

$$k_f = 1/(2.3 \times 0.2) = 2.2$$

That is, if eluted in isocratic chromatography with the solvent composition equal to that found at the time the peak left the column exit, the component will be expected to have a k' value of about 2.2. For very polar solutes which elute early in the gradient programme, this shortened form of the equation will not be valid and it is desirable to estimate the value of k_0 ($=k_w$). This can be found from rearranging equation 3.2 and substituting values for S and b:

$$k_w = 1/0.46 \times \{10^{[0.2(t_g - t_0 - t_d)/t_0]} - 1\} = k_0 \tag{3.7}$$

In any case it will be necessary to calculate a value for k_w for use in step (iv).

(iii) Calculate the solvent composition at the exit of the column at the time the peak of interest is eluted. This is found simply from

$$\%\ \text{MeOH} = \phi'(t_g - t_0 - t_d) \times 100 \tag{3.8}$$

That is, the exit solvent composition corresponds to the solvent composition that was being delivered at the peak retention time less the time delay corresponding to the column dead volume and the volume of any mixing chambers and connecting tubing, etc.

(iv) Calculate a new approximation for S from equation 3.4. In this calculation k' is set to the calculated value of k_f, the capacity factor expected in the exit solvent composition, which will be approximately 2.2.

(v) Calculate any desired k' value for a given solvent composition from

$$\phi_2 = 1/S\ (\log k_f - \log k') + \phi_{exit} \tag{3.9}$$

For a given column, the optimum k' value for maximum resolution under isocratic conditions is about 4 and, if this is satisfactory for the problem under consideration, all the previous equations become greatly simplified. What this means in practice is that an isocratic mobile phase corresponding to the mobile phase at time $t_g - 2.5\ t_0 - t_d$ is required. It may be that a k' value of 4 is not suitable for the application in mind and it will then be necessary to go through all the calculations. In fact, the calculations turn out to be easier than might first appear and are included as a BASIC program in Appendix II.

Turning again to the six-component mixture, a linear gradient separation is shown in Figure 3.2a. The value used for t_0 was 0.5 minute, requiring a 13.33% change in methanol concentration per minute (ϕ'), given $S=3$ and $b=0.2$ in equation (3.6). The gradient delay volume for the particular chromatograph used was found to be 0.6 ml. The gradient separation shown in Figure 3.2a shows only five components, perhaps indicating the care needed when interpreting a gradient separation. For this particular example it would be reasonable to position the third peak with a capacity factor (k') of about 4. In the gradient separation this third peak eluted at 6.11 min. Using the simple relationship described above, the isocratic composition required to give $k'=4$ for the third peak is $13.33 \times [6.11 - (2.5 \times 0.5) - 0.6]$ % methanol, i.e. 56.8% methanol. An isocratic separation using 56.8% methanol is shown in Figure 3.2b and the capacity factor of the third peak is 3.8, a perfectly acceptable prediction.

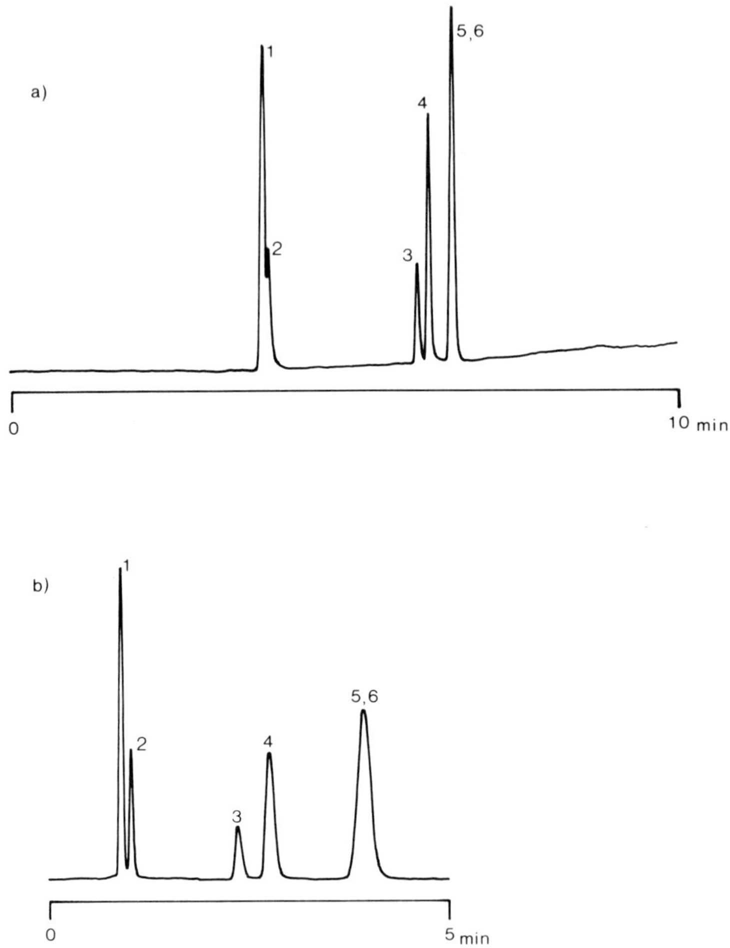

Figure 3.2. (a) Gradient separation of six-component test mixture using a constant rate of change of 13.3% methanol per minute; (b) isocratic separation using 56.8% methanol, estimated to give $k'=4$ for peak 3 from the initial gradient separation

There is no reason why the calculation need be restricted to just one peak; in fact, it is preferable to examine both the first- and last-eluted peaks. If it should turn out that the solvent compositions required to achieve reasonable retentions for these two peaks are widely divergent, then an isocratic separation will not be possible. To quantify this is difficult but one method would be to calculate the solvent composition required to give $k'=1$ for the first peak. This solvent composition is used in equation 3.4 to calculate the anticipated capacity factor for the last-detected peak. Should this capacity factor exceed a pre-specified limit, e.g. 10, then it can be concluded that it is unlikely that an isocratic separation can be achieved.

A similar procedure can be used for acetonitrile/water gradients, but here a value of 3–4 is more appropriate for S. In practice, the extrapolation from acetonitrile/water gradients to isocratic conditions has not always been found to be as reliable as when using methanol, so the latter solvent is to be preferred.

An alternative approach to predicting isocratic separations from gradient elution may be found in a paper by Schoenmakers *et al.*[25] Their aim was to improve the accuracy of the Snyder method by developing a graphical procedure based on exact mathematics. What they found was that the solvent strength parameter, S, was strongly correlated with k_0:

$$[S = p + q \ln k_0] \tag{3.10}$$

where p and q are constant for a given binary mobile phase system. If p and q are known, then retention in an isocratic system requires only one parameter to be known (S or $\ln k_0$).

For a linear gradient, the composition is a function of time, according to equation 3.5 and, as was shown above, the isocratic capacity factor is given by

$$[\ln k' = \ln k_0 - S\phi_b] \tag{3.4a}$$

If a sample is eluted before the gradient programme is completed, the retention time, t_R, is

$$\left[t_R = \frac{1}{S\phi'} \ln \left\{ 1 + S\phi' t_0 \exp \left[\frac{(1-\phi_0 q)S - p}{q} \right] \right\} \right] \tag{3.11}$$

For a sample eluted after the end of the gradient time (t_G), the retention time is

$$t_R = t_0 \exp \left[\frac{(1-qy)S-p}{q} \right] + \frac{1}{S\phi'} \left\{ \exp[S(\phi_0 - y)] - 1 \right\} + t_G \tag{3.12}$$

where y denotes the final composition. When used manually, a graphical method is employed in this scheme. However, for automated optimization, equations 3.11 and 3.12 must be solved numerically and values for p and q are required. For methanol/water systems, average values for p and q have been given as 2.86 and 0.77 for a 30 cm column having a void time of 125 seconds. For a different column geometry, flow rate or void time, new values of p and q must be calculated. This can be a tedious process since it requires that data are fitted to equation 3.4a to determine values for S which can then be plotted against k_0 values to provide estimates for the parameters p and q. Further guidance on the determination of values for p and q can be found in the work of Hafkenscheid and Tomlinson.[26] Comparing values of the two parameters from four sources they concluded that average values of 1.56 and 0.74, respectively, should be used.

Although this method was accurate and elegant in the hands of its designers, its transfer to other systems is not easy. Despite the approximations and limitations of the Snyder method, it must be preferred on account of its simplicity and ease of programming.

A third approach to conversion of gradient separations to isocratic separations, using acetonitrile as a modifier, was described briefly by Molnar.[27] The estimate of the isocratic compensation required to give retention times the same as those seen in the gradient separation is made from a calculation of the average composition of the mobile phase required to elute the particular solute. The average composition is found from

$$\phi_{AV} = (\phi_0 + \phi_R)/2 = \frac{\phi'}{2}(t_R - t_0) + \phi_0 \tag{3.13}$$

where ϕ_0 is the starting composition and ϕ_R is the composition at t_R, the peak retention time. Accuracies of better than 2% were claimed by Molnar for this method using linear acetonitrile/water gradients, but only one solute was used to test the approach. Additionally, equation 3.13 does not take into account any gradient delay volume.

Finally, another method based on calculating an average solvent composition has been used for both normal- and reversed-phase composition prediction.[8] The solvent composition (c_i) is calculated for each detected compound at the time it leaves the column from the generalized version of equation 3.8:

$$\phi = \phi'(t_g - t_0 - t_d) \tag{3.8a}$$

The required isocratic composition is found from the geometric mean of the solvent compositions c_i ($i = 1,2, \ldots, n$) at which the solutes i leave the column:

$$c = (c_1 c_2 c_3 \ldots c_i \ldots c_n)^{1/n} \tag{3.14}$$

Unfortunately, equation 3.14 usually over-estimated the solvent strength required and it was determined experimentally that by lowering the value by one quarter, a good estimate was obtained:

$$c_0 = 0.75c \tag{3.15}$$

The concentration used in the isocratic separation is therefore c_0. This method has been reported to be successful for methanol/water reversed-phase gradients and for dichloromethane/n-hexane gradients in normal-phase chromatography.

Gradient elution is a powerful method for the estimation of isocratic compositions but it is also a powerful separation method in its own right. The power and versatility of gradient methods are such that some simple methods of optimizing gradient profiles will be considered next.

3.4. GRADIENT ELUTION

In the context of this chapter the term *gradient elution* will be used to signify a separation which employs a mobile phase whose composition changes with time. For a binary mobile phase, the concentration of the stronger eluting solvent will be increased during the separation. The reasons for doing this are twofold. Firstly, gradient elution is an elegant way of solving the general elution problem: a scheme should be possible which permits all components to be eluted during a separation with peaks of very similar shape. This can be compared with most isocratic separations where early eluting peaks are narrow and often densely packed whereas later eluting peaks are very broad, often with excessively long retention times. The second reason is one already touched upon. Gradient elution with binary mobile phases, particularly in reversed-phase chromatography, provides a method of 'scouting' isocratic solvent compositions.

As indicated earlier, gradient elution has many problems associated with it and gradient separations can be more difficult than isocratic separations to optimize fully. Mobile phases may need to be especially purified and solvent compositions carefully matched, perhaps by the addition of inert (in a chromatographic sense) UV-absorbing agents such as nitrous oxide, to minimize effects on the detector baseline. There will be instrumental problems in the definition and repeatability of gradient profiles[20,21] which will restrict the utility and transferability of gradient methods. However, in spite of these drawbacks, gradient methods cannot be dismissed because they have so much versatility in the variety of possible gradient profiles.

To understand gradient elution chromatography fully, one should be familiar with the gradient elution equations. A number have been developed to describe the relationship between the observed solute retention time during a particular gradient and how the solute capacity factor may depend on the mobile phase composition. A succinct examination and interpretation of the more important equations has recently been presented.[28] The interest here is primarily in the application of the equations to the problems of optimizing gradient profiles.

As has already been described in Section 3.3, the starting point for a gradient separation should be a linear gradient. A linear change in solvent composition with time has the advantages of mathematical simplicity and ready interpretation. Linear gradients have been argued to provide fundamentally optimum separations based on a linear solvent strength (LSS) separation (see equation 3.1). For a typical set of solutes, the optimum slope for the gradient may be deduced from equation 3.6, i.e.

$$\phi' = b/St_0 \tag{3.16}$$

where ϕ' is the volume fraction change per unit time. Snyder *et al.*[14] suggested that b should have a value of 0.2 and S should be set to 3 for methanol, 3.1 for acetonitrile and 4.4 for tetrahydrofuran. Where high molecular weight solutes are being examined,[24] values for S should probably be increased by up to 100%.

Schoenmakers *et al.*[29] suggested that the linear log k' dependence described by Snyder should more correctly be a quadratic relationship, particularly for methanol and tetrahydrofuran gradients. The controversy over the behaviour was extended by Balke,[28] who argued that it is not possible to be definitive about which dependence description is absolutely correct since the statistical evaluations of data in both cases are inadequate. The controversy was further added to by an examination of the theoretical principles of Snyder's model.[30] Examining the separation of PTH-amino acids, good general agreement with the model was seen with a small number of specific exceptions. Thus it cannot be assumed *a priori* that LSS conditions will always be appropriate. However, the LSS method of Snyder is likely to provide a good starting gradient for further optimization.

A recommended starting point for a practical gradient separation[23] is to run a linear gradient of 5 to 95% acetonitrile/water. Pure water to pure organic modifier gradients can give rise to problems associated with eluent degassing and with the inability of many columns to operate efficiently with pure water as the eluent. The gradient slope can be estimated from equation 3.6, but a 'rule of thumb' value will be $(6.7/t_0)\%$ organic modifier per minute. The separation shown in Figure 3.2a testifies to the value of this simple rule. There are many ways in which this initial gradient can then be optimized. If resolution is inadequate, the gradient slope can be reduced or the column length increased. Optimization of the total analysis time is effected by varying the initial and final levels of the organic modifier. Flow rate could be used as a parameter in time optimization but care is needed since changes in flow rate, with all other factors held constant, can cause unexpected results. There are examples of peak order reversal in gradient elution as a consequence of flow changes.[31] In general terms the initial level of modifier should be increased until the first two peaks are just retained and resolved. The gradient slope and final conditions are then varied until the desired separation is achieved. If an adequate separation is not achieved, an alternative modifier should be sought.

An example of this approach to gradient optimization can be found in a paper by Lea.[24] Of course, it is not the only way to optimize gradient separations. Jandera and co-workers[32–34] provided a comprehensive insight into gradient theory and practice with good qualitative advice on gradient optimization, including a flow diagram for computer-aided gradient optimization.[33] Svoboda[35] described a computer approach to the search for an optimum gradient. The software itself is not listed but a useful flow chart of the algorithm is presented and the method is illustrated by the separation of nucleotides. A flow diagram of a computer program for optimizing gradient separations is also presented in a paper by Issaq *et al.*[36] The objective of the optimization is to provide a gradient profile (concave, linear or convex) which will give the same resolution for all the component pairs in the mixture. Other potential methods of optimizing gradient separations will also be considered in Chapters 4 (overlapping resolution mapping) and 6.

3.5. PRAGMATIC APPROACHES

The objective of the more pragmatic approaches is to try to predict the consequences of changes in experimental parameters having acquired a minimum amount of experimental data. To achieve this aim necessarily implies forcing experimentally derived data into theoretical models, accepting that there may be limitations. Ultimately, the aim will be to predict the separation based on a limited knowledge of the physico-chemical properties of the solutes being studied. In this section the possibility of separation prediction will be reviewed briefly. In practice, it is probable that insufficient information will be available to permit that prediction. The aim then will be to achieve a separation on the basis of minimum experimentation. If independent linear relationships can be assumed for the parameters of interest, then only two measurements will be required to investigate the influence of that parameter.

3.5.1. Solubility Parameters

The concept of solubility parameters arises from regular solution theory;[37] the solubility parameter is essentially a number which the chromatographer can use to indicate the required mode of chromatography and approximate retention characteristics. The solubility parameter (d) is defined as

$$d=(-E/v)^{1/2} \tag{16}$$

where E is the cohesive energy required to transfer 1 mole of the substance from the ideal gas phase into its liquid state (always negative) and v is the molar volume of the liquid. To give some idea of values, non-polar substances have d values around 7 $\mathrm{cal}^{1/2}\,\mathrm{cm}^{-3/2}$ (e.g. alkanes) while polar molecules, which exhibit much stronger liquid state interactions, have d values in excess of 10 $\mathrm{cal}^{1/2}\,\mathrm{cm}^{-3/2}$ (note alumina = 16) and up to 25 $\mathrm{cal}^{1/2}\,\mathrm{cm}^{-3/2}$ in the case of water.[38] The regular mixing rule can be used to describe activity coefficients in terms of solubility parameters and, since capacity factors can be expressed as ratios of activity coefficients, it follows that solubility parameters can be used to indicate chromatographic retention. Unfortunately, the successful use of tabulated values of solubility parameters for predicting retention has some way to go. Indeed, it has been suggested[38] that solubility parameters will find uses only in the qualitative or semi-quantitative description of retention and not in the prediction of retention. Nevertheless, good quantitative data have been generated over limited modifier ranges,[26] e.g. 35–75% methanol in water.

What emerges from the consideration of solubility parameters in HPLC is that solutes with similar solubility values, or polarities, will be difficult to separate (if their molar volumes are similar). The best chances of separating similar solutes comes from employing a chromatographic system with the highest solubility differences between the mobile phase and the stationary phase. The larger this difference, the more powerful the system will be for the separation of solutes with similar solubility parameters. For this reason it can be seen that reversed-phase chromatography (solubility differences of 10–15) is likely to provide greater selectivity than normal-phase chromatography (solubility differences of 5–10). To obtain reasonable retention times the polarity of the mobile phase should be adjusted so that the polarity of the solute molecule lies roughly mid-way between those of the stationary phase and the mobile phase. This is self-evident to most chromatographers, familiar with the need to use high water contents in the reversed-phase separation of carboxylic acids and other polar molecules, for example. The use of solubility parameters in automated optimization has still to be described, except that the theory has been used in the establishment of equations and models for use in other optimization methods.[39,40] Undoubtedly their consideration will be included in 'Expert Systems' now under development. On the other hand, solvophobic theory has been applied practically to automated optimization.

3.5.2. Solvophobic Theory

The 'solvophobic' theory of Horvath and co-workers[41–43] is a rigorous treatment of the solute distribution process in reversed-phase chromatography. It emphasizes the importance of the mobile phase, considering the exclusion of the less polar solute molecule from the polar mobile phase with subsequent sorption by the non-polar stationary phase. In other words, the solute is driven from the mobile phase rather than attracted to the stationary phase. It is to be expected that physico-chemical parameters such as partition coefficient, solute surface area and solubility (as seen above) will all play a part in determining the retention of a particular molecule and it should be possible to predict retention on the basis of quantitative structure–activity relationships. This relationship may be expected to have the form

$$\log k' = f_1X_1 + f_2X_2 + f_3X_3 \tag{3.17}$$

Such relationships have recently been established for alkylbenzenes and polyaromatic hydrocarbons,[44] substituted benzenes and phenols.[45,46] With such relationships, optimum separation conditions should be predictable using computer iteration, and this has indeed been shown to be the case. In ref. 46 a flow chart is given of a program which can automatically predict the optimum separation attainable in binary mobile phases (using methanol or acetonitrile as modifiers) as described by a chromatographic response function. Given that the system has the full physico-chemical data for the solutes of interest, it is necessary only to enter the solute names, the desired analysis time and the required resolution; everything else is automatic.

The solvophobic theory is of great interest and use in the development and understanding of chromatographic processes but still does not explain all retention phenomena. One of the problems with chemically bonded phases is that the nature of the phase is not constant. For example, the structure, and hence accessibility, of the long hydrocarbon chains of a C_{18} stationary phase will be quite different in mobile phases with a high water content in comparison with mobile phases with a high organic modifier content. In situations where very polar mobile phases are being used with stationary phases with long alkyl chains, the solute distribution more nearly resembles liquid–liquid partition. Solute retention now becomes dependent on the intrinsic stiffness of the hydrocarbon chains. A more detailed investigation into these effects, with the objective of providing a unified theory of chromatographic retention, was presented by Martire and Boehm[47] and no doubt the quest for a fuller understanding of retention mechanisms will continue until a good *a priori* estimation of separation conditions can be made. Until such time as this aim is realized, we shall have to continue with the more empirical approaches. In the next section the combination of empiricism with chromatographic theory provides methods for a more pragmatic approach to optimization.

3.5.3. Simple Semi-empirical Methods

Graphical methods must represent one of the simplest approaches to experimental optimization and it is usually a relatively simple matter to translate

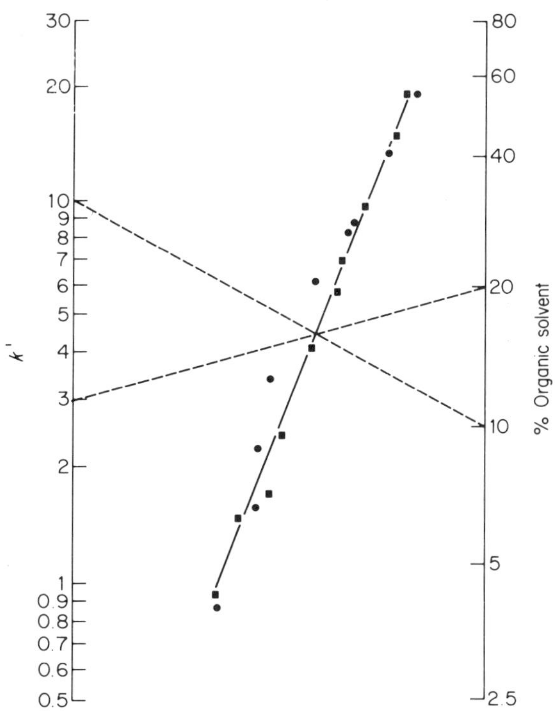

Figure 3.3. Nomogram for predicting effects of solvent composition on retention in reversed-phase chromatography. *Reprinted with permission from Ref. 48. Copyright (1980) American Chemical Society*

a graphical routine into a computer program. A nomogram (Figure 3.3) is available[48] for reversed-phase separations on octadecyl-bonded columns with high carbon loadings which enables retention in methanol- or acetonitrile-containing mobile phases to be predicted as a result of just a single experiment. The nomogram relies on the linear log k' versus composition relationship and will not be applicable to solutes which show marked non-linear dependences. The nomogram will probably be most useful for fine tuning of separations where deviations from linearity are less pronounced when only a small range of solvent composition is being considered. For automated use, a computer realization of the nomogram will be required, but this is relatively straightforward.

Another graphical method has been proposed[49] which uses plots of $\log(1 - t_0/t)$ (called R_Q) against composition. The validity of this transformation has been questioned[50] and the method cannot be recommended.

A problem with graphical methods is that they tend to be limited to the consideration of a single parameter, or possibly two, at a time because of the need to visualize and interpret them. In Chapter 5 the use of 'window diagrams' as a graphical method for optimization will be discussed, where more complex inter-relationships can be considered. Computers, fortunately, do not suffer from such visual limitations. Even a programmable calculator can be very effective. Software was published for a Texas Instruments TI 59 calculator which was intended to allow the chromatographer to investigate the effects of temperature and composition for methanol/water reversed-phase separations.[51] Unfortunately, the program listing as given is incorrect and cannot be used; a corrected listing was not published.[52] The approach is of interest and has been extended to acetonitrile/water separations,[53] again with software for a TI 59, but on this occasion the software was not published. The optimization process is based on the following well known resolution equation:

$$R = 2(t_2 - t_1)/(w_1 + w_2) \tag{3.18}$$

where t is the peak retention time and w the peak base width. The resolution R is also related to capacity factors (k') and the column plate number (N):

$$t = t_0(1 + k') \tag{3.19}$$

$$N = 16(t/w)^2 \tag{3.20}$$

The plate number, N, can be found as a function of experimental conditions using equations (3.21) and (3.22):

$$h = 2/\nu + A\nu^{0.33} + C\nu \tag{3.21}$$

$$h = H/d_P \tag{3.22}$$

where h is the reduced plate height,[54] ν is the reduced velocity of the mobile phase and d_P is the particle diameter of the stationary phase. For modern porous packings in well packed columns, values of 2 and 0.05 can be used for A and C, respectively.

The reduced velocity of the mobile phase, ν, is given by

$$\nu = ud_{\mathrm{P}}/D_{\mathrm{m}} \tag{3.23}$$

where u is the linear velocity of the mobile phase. The value of u is simply L/t_0, the column length divided by the void time. D_{m} is more difficult to calculate; it is the solute diffusion coefficient, commonly estimated using the Wilkie–Chang equation.[55] To calculate this requires a knowledge of how the mobile phase viscosity changes with temperature. Empirical equations are available for methanol/water[51] and acetonitrile/water[53] mobile phases.

The mobile phase velocity is also a function of the inlet pressure (P), the column permeability ϕ^* and mobile phase viscosity (η):

$$u = Pd_{\mathrm{P}}^2/\phi^*\eta L \tag{3.24}$$

The dependence of capacity factors on absolute temperature (T) can be described by

$$\log k'_2 = \log k'_1 - a(1/T_1 - 1/T_2) \tag{3.25}$$

The dependence of capacity factors on composition can be assumed to follow the now familiar logarithmic form (e.g. equation 3.4).

The actual mathematics are relatively straightforward and are presented as a BASIC program in Appendix III. The experimental data needed are simply retention data under four different conditions, two compositions at each of two temperatures. This method permits interactive optimization, the analyst being able to explore the effects of composition and temperature changes on resolution, time, inlet pressure, etc. The limitations are that it considers just two peaks at a time and that only two modifiers are catered for.

It is possible to extend this simple approach to more modifiers and to normal-phase chromatography.[56,57] Again, linear relationships are assumed for dependences of the capacity factor on all the variables to be optimized (flow rate, temperature, mobile phase composition, column length and particle diameter). Additionally, this extended pragmatic approach is also an interactive method, allowing the analyst to specify and examine particular constraints such as time or minimum resolution, to obtain simulated chromatograms and to adjust the separation on the basis of local criteria such as costs or sensitivity. A computer program has been described[56,57] which permits the estimation of analysis time, resolution, inlet pressure and peak height as functions of six separation parameters. The parameters are column length, particle diameter, mobile phase velocity, up to two modifiers and temperature. A maximum of seven experiments for each compound is required, but this can still lead to large numbers of chromatograms for complex mixtures. Fortunately, the computer program designs the experimental matrix appropriate to the problem being considered. Unfortunately, it is possible to do no more than whet appetites to this approach, since a listing of the software is not, as now seems to be the accepted practice, contained in the paper.

By forcing the pragmatic approach, it becomes possible to confer upon computer-controlled optimization an impression of 'artificial intelligence'. Just such an approach has been described[58] where general equations were derived from statistical thermodynamics and subsequently used to predict retention behaviour in both normal- and reversed-phase chromatography. However, experimental data are required for parameter determination (three or four experiments), the solute peaks need to be identified in successive chromatograms and much of the intelligence required is far from artificial!

3.6. CONCLUSIONS

The methods presented in this chapter are worthy of further exploration for those interested in a quick, although possibly limited, answer and for those not possessing 'state of the art' instrumentation. Difficulties over the initial choice of a column can be reduced by selecting just one (with a cyano-bonded phase) and then experimenting with a selected range of mobile phases covering both normal- and reversed-phase modes. The general elution problem can be overcome by prudent use of gradients if solvents can be obtained which are sufficiently clean. The advantages of trying gradient separations are that they can give information about conditions necessary for isocratic separations and whether an isocratic separation will be possible.

The inherent power of gradient elution can make its optimization difficult but, particularly for binary gradients, the first step will be a linear gradient with a slope defined by the column dimensions and flow rate. Further optimization of the gradient profile can be restricted to adjustment of initial and final compositions.

Chromatographic theory continues to develop such that an estimate of separation conditions can be made on the basis of limited physico-chemical information on the solutes. Solubility and solvophobic theories are contributing much to this area, but cannot help initially in the case of unknown mixtures.

The more pragmatic approaches are becoming increasingly powerful. Their widespread use is currently limited by the shortage of software, a situation that is unlikely to continue for long.

Many of the procedures and approaches use basic and fundamental chromatographic equations which should be familiar to all. Are these methods sufficient in their own right? With the exception of the more extensive pragmatic approaches, probably they are not. These methods do, however, constitute a set of ideas which can form the basis, or starting point, for a more detailed and comprehensive optimization utilizing one of the approaches contained in the following chapters.

3.7. REFERENCES

1. D. L. Massart and G. Hoogewijs, *Pure Appl. Chem.*, **55** (1983) 1861.
2. M. R. Detaevernier, G. Hoogewijs and D. L. Massart, *J. Pharm. Biomed. Anal.*, **1** (1983) 331.
3. G. Hoogewijs and D. L. Massart, *J. Pharm. Biomed. Anal.*, **1** (1983) 321.
4. G. Hoogewijs and D. L. Massart, *J. Pharm. Biomed. Anal.*, **2** (1984) 449.
5. G. Hoogewijs and D. L. Massart, *J. Liq. Chromatogr.*, **6** (1983) 2521.
6. G. Hoogewijs and D. L. Massart, *J. Pharm. Belg.*, **38** (1983) 76.
7. G. Hoogewijs and D. L. Massart, *J. Chromatogr.*, **309** (1984) 329.
8. M. de Smet, G. Hoogewijs, M. Puttemans and D. L. Massart, *Anal. Chem.*, **56** (1984) 2662.
9. L. Sybrandt and E. Montoya, *Am. Lab.*, **9**(8) (1977) 79.
10. F. W. Karasek, *Res. Dev.*, **28**(6) (1977) 38.
11. V. V. Berry, *J. Chromatogr.*, **199** (1980) 219.
12. S. M. McCown, B. E. Morrison and D. L. Southern, *Int. Lab.* **14**(5) (1984) 76.
13. V. V. Berry, *J. Chromatogr.*, **236** (1982) 279.
14. V. V. Berry and R. E. Shansky, *J. Liq. Chromatogr.*, **7** (1984) 943.
15. V. V. Berry and R. E. Shansky, *J. Chromatogr.*, **284** (1984) 303.
16. S. Yamamoto, K. Nakamaru and Y. Morikawa, *J. Liq. Chromatogr.*, **7** (1984) 1033.
17. V. V. Berry, *J. Chromatogr.*, **290** (1984) 143.
18. Sj. Van Der Wal and L. R. Synder, *J. Chromatogr.*, **255**, (1983) 463.
19. V. V. Berry, J. Chromatogr., 321 (1985) 33.
20. M. A. Quarry, R. L. Grob and L. R. Snyder, *J. Chromatogr.*, **285** (1984) 1.
21. M. A. Quarry, R. L. Grob and L. R. Synder, *J. Chromatogr.*, **285** (1984) 19.
22. L. R. Snyder, J. W. Dolan and J. R. Gant, *J. Chromatogr.*, **165** (1979) 3.
23. J. W. Dolan, J. R. Gant and L. R. Snyder, *J. Chromatogr.*, **165** (1979) 31.
24. A. G. H. Lea, *J. Chromatogr.*, **194** (1980) 62.
25. P. J. Schoenmakers, H. A. H. Billiet and L. de Galan, *J. Chromatogr.*, **205** (1981) 13.
26. T. L. Hafkenscheid and E. Tomlinson, *J. Chromatogr.*, **264** (1983) 47.
27. I. Molnar, *J. High Resolut. Chromatogr. Chromatogr. Commun.*, **4** (1981) 276.
28. S. T. Balke, *Quantitative Column Liquid Chromatography, A Survey of Chemometric Methods*, Journal of Chromatography Library, Vol. 29, Elsevier, Amsterdam, 1984.
29. P. J. Schoenmakers, H. A. H. Billiet and L. de Galan, *J. Chromatogr.*, **185** (1979) 179.
30. K. A. Cohen, J. W. Dolan and S. A. Grillo, *J. Chromatogr.*, **316** (1984) 359.
31. J. L. Glajch, M. A. Quarry and L. R. Snyder, 8th International Symposium on Column Liquid Chromatography, New York, 1984, paper 1p–B2.
32. P. Jandera, M. Janderova and J. Churacek, *J. Chromatogr.*, **192** (1980) 19.
33. P. Jandera and J. Churacek, *Adv. Chromatogr.*, **19** (1981) 125.
34. P. Jandera and J. Churacek, *Gradient Elution in Column Liquid Chromatography, Theory and Practice*, Elsevier, Amsterdam, 1985.
35. V. Svoboda, *Radioisotopy*, **18** (1977) 775.
36. H. J. Issaq, K. L. McNitt and N. Goldgaber, *J. Liq. Chromatogr.*, **7** (1984) 2535.
37. J. H. Hilderbrand, J. M. Prausnitz and R. L. Scott, *Regular and Related Solutions*, Van Nostrand Reinhold, New York, 1970.
38. P. J. Schoenmakers, H. A. H. Billiet and L. de Galan, *Chromatographia,* **15** (1982) 205.
39. P. J. Schoenmakers, *J. Chromatogr.*, **315** (1984) 1.
40. P. J. Schoenmakers, H. A. H. Billiet and L. de Galan, *J. Chromatogr.*, **282** (1983) 107.
41. Cs. Horvath, W. Melander and I. Molnar, *J. Chromatogr.*, **125** (1976) 129.

42. Cs. Horvath and W. Melander, *Am. Lab.*, **10**(10) (1978) 17.
43. W. Melander, J. Stoveken and Cs. Horvath, *J. Chromatogr.*, **199** (1980) 35.
44. M. N. Hasan and P. C. Jurs, *Anal. Chem.*, **55** (1983) 263.
45. K. Jinno and K. Kawasaki, *J. Chromatogr.*, **298** (1984) 326.
46. K. Jinno and K. Kawasaki, *J. Chromatogr.*, **316** (1984) 1.
47. D. E. Martire and R. E. Boehm, *J. Phys. Chem.*, **87** (1983) 1045.
48. J. L. Meek, *Anal. Chem.*, **52** (1980) 1370.
49. S. Toon and M. Rowland, *J. Chromatogr.*, **208** (1981) 391.
50. E. Tomlinson, *J. Chromatogr.*, **236** (1982) 258.
51. J. R. Gant, J. W. Dolan and L. R. Snyder, *J. Chromatogr.*, **185** (1979) 153.
52. J. R. Gant, J. W. Dolan and L. R. Snyder, *J. Chromatogr.*, **192** (1980) 504 (erratum).
53. J. C. Berridge, *J. Chromatogr.*, **202** (1980) 469.
54. J. H. Knox, *J. Chromatogr. Sci.*, **15** (1977) 352.
55. B. L. Karger, L. R. Snyder and C. S. Horvath, *An introduction to Separation Science*, Wiley Interscience, New York, 1973, p. 78.
56. J. P. Bounine, G. Guiochon and H. Colin, *J. Chromatogr.*, **298** (1984) 1.
57. J. P. Bounine and G. Guiochon, *Analusis*, **12** (1984) 175.
58. L. Peichang and L. Xiaoming, *J. Chromatogr.*, **292** (1984) 169.

4 The Solvent Selectivity Triangle and Mixture Designs

4.1. INTRODUCTION

One of the major problems in drawing together the many techniques available for HPLC separation optimization in a book such as this is their classification. Within this chapter will be considered one of the largest groups of techniques that are available to the chromatographer. What links them together is the way in which the optimization problem is approached. All the methods begin by requiring that retention data are measured in a part of a multi-variable space that is defined by the various parameters under consideration. The positions at which the data are obtained are decided by a formal experimental design. Furthermore, many, if not all, experiments must be completed before any data analysis can be carried out. For this reason the methods are sometimes referred to as 'simultaneous' techniques (as opposed to 'sequential' techniques—see Chapter 6), although this does not mean that all experiments are carried out at exactly the same time! When all the measurements have been accumulated, the data are analysed and fitted to a model from which it is hoped the optimum separation conditions will be apparent. This provides a second classification for the methods—they are model dependent. The use of a formal experimental design simplifies the fitting of the experimental data to the model. By using a computer method to calculate an optimization criterion, to plot a map of the separation quality or to conduct a detailed search, the optimum separation conditions should be predictable.

In this chapter the methods are discussed only in the context of chromatographic optimization but, as will be seen in Chapters 5 and 6, there are other formal experimental designs of a more general nature which can be very appropriately applied to HPLC. Indeed, many use some of the ideas and concepts that will be described in this chapter—there is considerable overlap in methods and concepts.

A second common thread that links many of these methods is contained in the solvent classification known as 'the solvent selectivity triangle', first proposed by Snyder,[1,2] which encompasses both reversed- and normal-phase chromatography. There are a number of approaches which have been developed to exploit this classification and there is little doubt that this formal solvent classification has had a dramatic impact on the way separation development is approached.

The first optimization technique to be considered will be factorial designs. While not restricted to chromatography, the philosophy it embraces has so much in common with methods that follow that it is useful to include it here. Factorial designs are ideally suited to the optimization of the more discrete variables such as pH or to an initial 'scouting' of the factor space to limit the area of interest.

The mapping method, more generally described as 'overlapping resolution mapping' (ORM), for predicting optimum chromatographic conditions was first described by Glajch *et al.*[3] and it is an extremely powerful method provided that certain minimum information is available. The power and capabilities of this method, for isocratic and gradient reversed- and normal-phase separations, warrant a detailed consideration. Its developments towards the optimization of the more discrete variables, such as column stationary phase and the pH of the mobile phase, will also be described.

Finally, a more simplistic approach will be considered briefly. In this case the region of experimental interest is first defined and then the whole of that area is searched in a step-by-step manner. No decisions are made about the quality of the separations achieved until all the data have been accumulated. The merits of a simple approach lie in its ease of use and applicability to any situation. The reasons for including it in this chapter are that it shares similarities with the overlapping resolution mapping method since it presents a graphical picture of regions worthy of further investigation. It is also interesting to consider the relative merits of a simplistic approach with the more esoteric ORM method.

4.2. THE SOLVENT SELECTIVITY TRIANGLE

The phenomenal success of HPLC as a versatile analytical tool surely stems from the intrinsic selectivity that is attainable from the use of mixed solvent systems. Practically all optimization methods require that a decision is first made about which few solvents will be selected for the method to explore. Since the successful exploitation of the methods about to be described owes so much to the rational classification of solvent properties, it is appropriate to consider this in some detail.

Referring to the well known resolution equation for HPLC:

$$R = \tfrac{1}{4}(\alpha - 1)N^{1/2}k'/(1 + k') \qquad (4.1)$$

it becomes evident that the primary interest is in the α term, the selectivity factor. The column parameters are usually constant, giving a fixed value to the plate number N, and peaks are normally arranged to elute with capacity factors (k') in the range 1–10. A major objective in separation optimization is to make the value of α as large as possible, consistent with other constraints that may be placed on the separation, such as time and detection sensitivity. The α term is influenced more by solvent composition than by the solvent strength, and this will be self-evident to most chromatographers–we intuitively explore ternary and even quaternary mobile phases when faced with a difficult separation. The relative ability of chemically different solvents to interact with the various compounds that are to be separated is the property that is most likely to affect changes in α values, although the contribution from different stationary phases should not be ignored.

The empirical quantitative relationship between strength and selectivity was proposed by Rohrschneider[4] and further developed by Snyder,[1] about 80 different solvents being considered. The solvent strength (often referred to as polarity) can be an approximate measure of the ability of a solvent to elute a compound from a chromatographic column, and this measure has been placed on a scale of P' values (see Table 4.1). These values can be used to enable a rough prediction of the selection of solvent(s) for the optimization of retention. The relative abilities of each particular solvent to interact with the solute of

TABLE 4.1. Suggested solvents for HPLC (from ref. 1)

	P'	x_e	x_d	x_n	Solvent group
Normal-phase:					
Diethyl ether	2.8	0.53	0.13	0.34	I
Di-n-butyl ether	2.1	0.44	0.18	0.38	I
Chloroform	4.1	0.25	0.41	0.33	VIII
Dichloromethane	3.1	0.29	0.18	0.53	V
Ethyl acetate	4.4	0.34	0.23	0.43	VI
Isopropanol	3.9	0.55	0.19	0.27	II
n-Propanol	4.0	0.54	0.19	0.27	II
Ethanol	4.3	0.52	0.19	0.29	II
Acetonitrile	5.8	0.31	0.27	0.47	VI
Water	10.2	0.37	0.37	0.25	VIII
n-Hexane	0.1				Carrier
Isoctane	0.1				Carrier
Reversed-phase:					
Methanol	5.1	0.48	0.22	0.31	II
Tetrahydrofuran	4.0	0.38	0.20	0.42	III
Acetonitrile	5.8	0.31	0.27	0.47	VI
Dioxane	4.8	0.36	0.24	0.40	VI
Dimethylformamide	6.4	0.39	0.21	0.40	III
Methoxyethanol	5.5	0.38	0.24	0.38	III
Water	10.2				Carrier

interest can be measured by x_d (proton donor), x_e (proton acceptor) and x_n (dipole–dipole) values. Reference to these values will be of interest and use in the selection of solvents which will show the different chemical interactions, and hence selectivities, with the sample to be separated. Table 4.1 lists the four parameters just described for the solvents most likely to be of use in most HPLC applications. Solvents with the widest potential differences in their interactions will be expected to be most successful in producing a separation of high selectivity.

These three major mobile phase effects which contribute to the selectivity result from proton donor, proton acceptor and dipole–dipole interactions with the compounds to be separated. The relative selectivities of chromatographic solvents have been very conveniently grouped into sets by Snyder,[1] as can be deduced from Table 4.1. Figure 4.1 is a plot of the different selectivity groups of solvents, using trilinear coordinates, of their proton donor, proton acceptor and dipole–dipole interactions. Solvents which are separated by the classification can be expected to provide the greatest scope for selectivity effects in chromatography. Each circle on the plot embraces a group of solvents with similar selectivity characteristics and generally the same functionality: this can be seen from the groups of solvents listed in Table 4.2.

Using this solvent classification, it is possible to develop a strategy for the prior selection of solvents for mobile phase optimization. To effect changes in selectivity, solvents from selectivity groups as far apart as possible should be

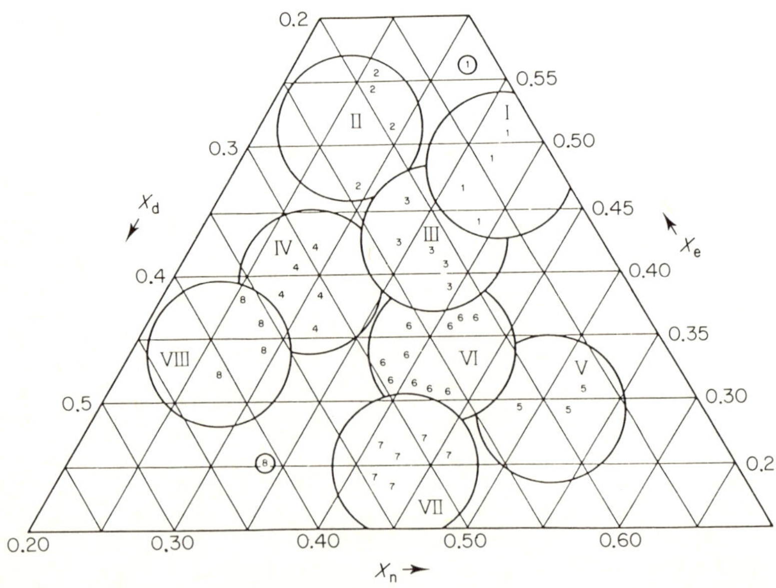

Figure 4.1. Selectivity grouping of solvents according to Table 4.1. *Reproduced from the* Journal of Chromatographic Science, *ref. 1, by permission of Preston Publications Inc*

TABLE 4.2. Classification of solvent selectivities

Group	Solvents
I	Aliphatic ethers, tetramethylguanidine, hexamethylphosphoramide (trialkylamines)
II	Aliphatic alcohols
III	Pyridine derivatives, tetrahydrofuran, amides (except formamide), glycol ethers, sulphoxides
IV	Glycols, benzyl alcohol, acetic acid, formamide
V	Dichloromethane, dichloroethane
VI	(a) Tricresyl phosphate, aliphatic ketones and esters, polyethers, dioxane
	(b) Sulphones, nitriles, propylene carbonate
VII	Aromatic hydrocarbons, halogen-substituted aromatic hydrocarbons, nitro compounds, aromatic ethers
VIII	Fluoroalkanols, *m*-cresol, water (chloroform)

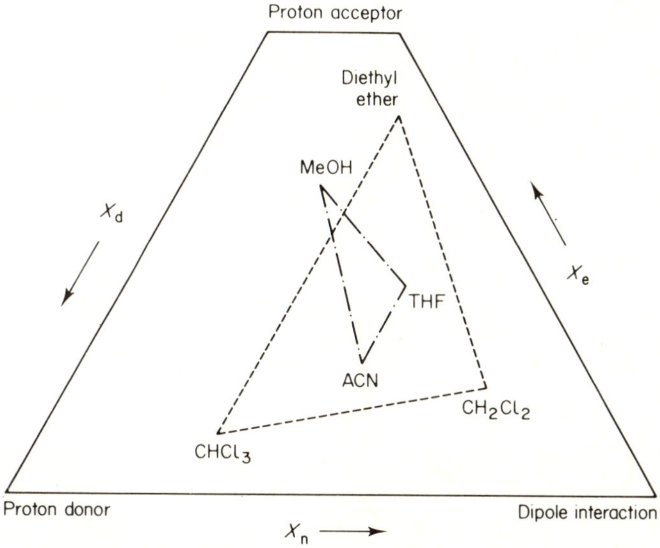

Figure 4.2. Selectivity triangles for the preferred solvents in reversed-phase (— · — · —) and normal-phase (-----) chromatography. *Reproduced from ref. 3 by permission of Elsevier Science Publishers BV*

chosen. Since there are three variables in the classification of the solvent properties, the solvents can be considered as forming the corners of a triangle—a 'solvent selectivity triangle'. By making this triangle as large as possible, the maximum contrast in solvent properties will be achieved and this will provide the greatest opportunity to produce the desired selectivity. It is not practically possible to choose solvents from groups at the extreme vertices of the classification shown in Figure 4.1 on the grounds of cost, chromatographic suitability (e.g. the solvent may absorb UV light) or miscibility with the other solvents. Selection of the two most useful sets of solvents, one for normal- and one for reversed-phase chromatography, produces two solvent selectivity triangles. These two triangles are depicted in Figure 4.2, using the preferred solvents for the respective modes to define the vertices of the triangles. For normal-phase chromatography the solvents would be diethyl ether (or, more preferably, on account of its greater stability and lower volatility methyl *tert*-butyl ether), dichloromethane and chloroform. These solvents would be combined in various proportions to manipulate selectivity and would be mixed with n-hexane as diluent for adjustment of overall solvent strength. The analogous solvents for reversed-phase chromatography would be methanol, tetrahydrofuran and acetonitrile with water for solvent strength adjustment.

One should not, however, be mislead into assuming that separations should be attempted with only these solvents, since it may be appropriate to select a different solvent to achieve the desired selectivity. If a microbore column is in use, it will be feasible to consider some of the more unusual and expensive solvents to provide an otherwise unattainable selectivity. Note that this classification does not seek to exploit the properties of other solvents which are added in small amounts to control pH or which can be used to improve the efficiency of separations by modifying the nature of the stationary phase. In this latter context can be included aryl- and alkylamines (often used to mask residual silanol effects) and solvents such as dimethylformamide.[5] One should also be aware that there will be exceptions that do not conform to this classification. The elution order of isomeric trimers of polystyrene has been found[6] to be independent of the dominant solvent–solute interactions as defined in the solvent selectivity triangle. Further, where standard silica or hydrocarbon-bonded supports are not being used, specific interactions may determine the effective solvent strength. The selectivity of carbon adsorbents is completely different to nonpolar chemically bonded phases, which implies also a different consideration of solvent strength[7] and the range of modifiers which could prove useful.[8]

While the above classification has been found to be generally appropriate, particularly in reversed-phase chromatography, some limitations in normal phase liquid–solid chromatography (LSC) have been noted.[9] There is little argument that in bonded-phase chromatography the solvent selectivity triangle provides a rational guide for the selection of solvents which might be expected to provide the maximum selectivities. For many LSC applications it has been demonstrated that *localization* effects in the stationary phase are of critical importance.[10] These localization effects result from the competition between mobile phase and solute molecules for a position directly over the adsorption site of the stationary phase. For silica supports these sites would be the silanol groups.

The localization effects can be subdivided into solvent–solute localization and solvent-specific localization. The degree of solvent localization can be measured by a mobile phase parameter m, which increases as the concentration of the polar component of the mobile phase increases. Two solvents that demonstrate this effect are methyl *tert*-butyl ether and acetonitrile, both of which give large values of m for mobile phases in which they are included. Solvent-specific localization is deduced to involve the direct hydrogen bonding of a basic, polar solvent with active surface sites on the stationary phase. For example, basic polar solvents such as methyl *tert*-butyl ether can be expected to exert different and additional selectivity effects when compared with less basic but still polar solvents such as acetonitrile. This feature then gives the ability to adjust selectivity by changing the ratios of such solvents while still maintaining the overall value of m approximately constant.

Consideration of the implications of these solvent localization effects led Glajch *et al.*[10] to propose an alternative solvent selectivity triangle for liquid–solid chromatography. This triangle is still directly analogous to the triangle proposed for reversed-phase chromatography (and to some extent that previously recommended for normal-phase chromatography) in that the corners correspond to the solvents expected to provide the extremes in selectivity. For the 'solvent localization triangle' the vertices of the triangle consist of a non-localizing (NL) solvent such as dichloromethane, a basic, localizing solvent (LB) such as methyl *tert*-butyl ether and a non-basic localizing (localizing dipole, LD) solvent such as acetonitrile. By using mixtures of these three solvents, a wide range of selectivities should be attainable. To control the total retention a fourth solvent, one not expected to make a significant contribution to the overall selectivity, is included. For LSC the choice here would be n-hexane, isooctane or 1,1,2-trichloro-1,2,2-trifluoroethane (Freon-113).[11]

The existence of two solvent selectivity triangles from which to choose in normal-phase chromatography does not simplify the selection of starting conditions. Perhaps much still remains to be elucidated for this mode of chromatography. Certainly, looking at the literature on normal-phase separations, many 'odd' solvents have been employed. The position of compromise would be to use dichloromethane and methyl *tert*-butyl ether as the two primary modifying solvents, with n-hexane or Freon-113 as the carrier (note that the Freon has the advantage of greater miscibility, particularly with solvents such as acetonitrile). The third solvent can be chosen from acetonitrile and chloroform. Acetonitrile may provide the higher selectivity potential since it represents a polar, non-basic solvent with a large mobile phase localization selectivity. Unfortunately, acetonitrile is not freely miscible with n-hexane but if Freon-113 is used as the independent solvent strength adjuster then all compositions are available. Of course, chloroform may not be suitable because of its comparatively high UV cut-off wavelength. Further theoretical considerations and practical guidance on the selection of solvents in liquid–solid chromatography are available in an excellent review.[12]

The liquid–solid localization classification is not restricted to use with silica columns alone. It has been found[13] that retention of less polar solutes on cyanopropyl-silica columns can be quantitatively correlated with the localization model. It appears that fully bonded cyanopropyl-silica columns behave as deactivated silica columns when moderately polar or nonpolar mobile phases are used. In other words, the adsorption sites are hindered silanol groups rather than cyanopropyl moieties. Method development and separation optimization using cyanopropyl-silica columns should, therefore, be approachable in the same way as for unmodified silica columns. This does indeed seem to be the case as independent studies[14,15] have shown very similar solvent strengths and selectivities for cyanopropyl and silica columns. Aminopropyl-modified silica columns were found to behave differently[15] and offer selectivity advantages.

The classifications for normal-phase chromatography described above by no means represent all the classifications that can be considered. For example, a rational method for the selection of solvents based on their having isohydric (i.e. equal water) properties[16] has been proposed. Thus an initial investigation of the two ternary phases A and B described below is recommended. It should be noted that an acid or base is added, depending on the solutes under investigation.

Mobile phase A: isooctane/diisopropyl oxide/acetic acid (or triethylamine)/ water (50:50:0.2:0.0041)

Mobile phase B: diisopropyl oxide/methanol/acetic acid (or triethylamine)/ water (50:50:0.2:2.6)

The proportions of the two phases A and B are adjusted to achieve the desired retention of the solutes. The advocates of this method estimate that more than 90% of normal-phase separations could be achieved with these two mobile phases and mixtures thereof. It will be seen, however, that in many ways the method here is a subset of the triangle classification. The adjustment of the water content, to maintain a constant water activity of the support, is interesting. This approach can be extremely valuable in aiding reproducibility and permitting rapid changes from one mobile phase to another, as time for water equilibration of the silica is not required.

The main practical results that come from a systematic classification of solvents are that the classification promotes an easier selection of a group of solvents which can be explored in a more logical manner. It will not always be obvious which particular solvents or additives are providing the greatest contributions to the overall selectivity of the separation. Such information can be obtained from the analysis of suitable formal experimental designs. Factorial designs represent such formal experimental protocols designed to aid the identification of the primary and secondary effects of the variables under investigation.

4.3. FACTORIAL DESIGNS

4.3.1. Theory

It has been suggested[17] that, when considering the strategies available for chromatographic optimization, the variables involved can be conveniently grouped into two classifications, *discrete* and *related* variables. The individual components of the mobile phase would be considered to be related variables as they directly affect each other—one cannot explore 0–100% methanol and simultaneously explore 0–100% tetrahydrofuran since the total composition of the mobile phase must obviously remain at 100%. On the other hand, temperature and pH might be considered to be discrete variables because, in themselves, they can be freely varied without directly causing any changes to the values of other parameters. It should be made clear, though, that this classification does not imply that the discrete variables are independent of each other. If the effect of one variable (or factor) depends on the value of another factor, then the factors are said to interact mathematically. In chromatography this is the rule rather than the exception. Perhaps one of the points of interest is to establish which parameters are having a fundamental influence on the chromatography and just how interdependent the variables actually are for the separation under study. This is where use of formal statistical experimental designs can be valuable.

The subject of statistical experimental design is extremely large and varied. There are many sources which provide excellent introductions and practical guidance to the many aspects available (e.g. ref. 18). *Factorial designs* are especially useful for discovering factor effects and their interactions. The applications of factorial designs, particularly in analysis, are many and their wider uses are also well described in more general texts.[18–21] Perhaps the most useful feature of factorial designs is that the experiments are designed to provide data which are ideally suited to regression analysis and model fitting. The experimental design can also be randomized such that hidden effects do not compromise the ultimate results. Such hidden effects arise from uncontrolled variables which, for HPLC, might include the overall performance of the column in use, which may be deteriorating in an unknown manner throughout the experimental period. Without randomizing an effect such as this it might appear that a whole region of the experimental factor space is ruled out; in fact, all that may have occurred is that the column performance deteriorated to an unacceptable level when these experiments were conducted. Factorial designs also have the great advantage of *initially* assuming that each variable behaves in an independent manner. In other words the design assumes that the importance of each individual variable can be clearly seen. In HPLC, the variables are not usually totally independent. The subsequent analysis of the data generated from a factorial design clearly shows whether there is independence or whether interaction occurs between combinations of the variables. Finally, for a full exploration of the influences of the variables, a factorial design provides a highly efficient method of obtaining the necessary data with the required degree of precision and minimum redundancy of effort.

There are, unfortunately, some negative consequences of using factorial designs. The positions at which experiments should be conducted needs some careful forethought. If experiments are set too far apart, the results may indicate that no change has taken place. This result can be experienced particularly with discrete variables such as exist in mobile phase optimization and in situations where peak crossover occurs. The area of interest may be in the mid-range of the experiments and may be missed by a standard factorial design. As will be seen, there are experimental designs, such as the central composite design, which overcome to some extent the problems associated with missing information at the central part of the factor space. The other side of that coin is if the experiments are placed too close together, any effects that are occurring may be insignificant in relation to the experimental noise and errors. It may certainly pay dividends to precede a factorial optimization by some range-finding experiments so that the size of the design can be appropriately assessed. Another problem associated with factorial designs and all analogous methods is that all the data are required prior to analysis. If the full design cannot be completed, the data so far accumulated may be extremely difficult to analyse. This should be borne in mind before setting off on a multifactor, multilevel design!

There are many variants of factorial designs but the basic procedure is to carry out experiments where each variable is investigated at two levels. Each variable is given two values, an upper (+) and a lower (−), which then define the limits of the experimental domain (often referred to as the 'factor space'). This is depicted in Figure 4.3 and Table 4.3 for three variables at two levels. It can be seen that eight experiments will be required to define the factor space. In general, for n variables, 2^n experiments will be needed. With a two-level factorial design it is assumed that linear equations can be used to describe the effects of the variables. In liquid chromatography this is seldom the case. To fit a higher order model (e.g. a second-order polynomial) more data points are

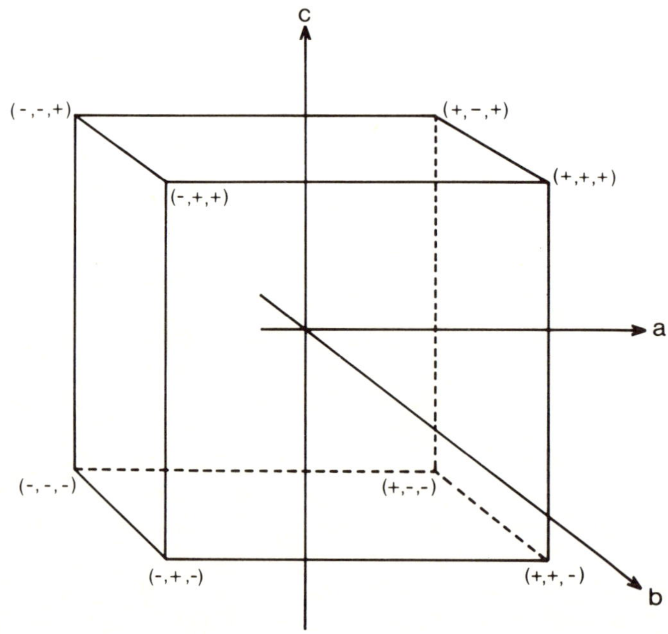

Figure 4.3. Full factorial design for three variables (a,b and c): experiments are carried out at each corner of the cube

TABLE 4.3. Factorial design for three variables at two levels

Experiment No.	A	B	C	Results
1	+	+	+	Y_1
2	+	+	−	Y_2
3	+	−	+	Y_3
4	+	−	−	Y_4
5	−	+	+	Y_5
6	−	+	−	Y_6
7	−	−	+	Y_7
8	−	−	−	Y_8

Effects

$A = \frac{1}{4}[(Y_1 + Y_2 + Y_3 + Y_4) - (Y_5 + Y_6 + Y_7 + Y_8)]$
$B = \frac{1}{4}[(Y_1 + Y_2 + Y_5 + Y_6) - (Y_3 + Y_4 + Y_7 + Y_8)]$
$C = \frac{1}{4}[(Y_1 + Y_3 + Y_5 + Y_7) - (Y_2 + Y_4 + Y_6 + Y_8)]$

Interactions

$A \times B = \frac{1}{4}[(Y_1 + Y_2 + Y_7 + Y_8) - (Y_3 + Y_4 + Y_5 + Y_6)]$
$B \times C = \frac{1}{4}[(Y_1 + Y_4 + Y_5 + Y_8) - (Y_2 + Y_3 + Y_6 + Y_7)]$

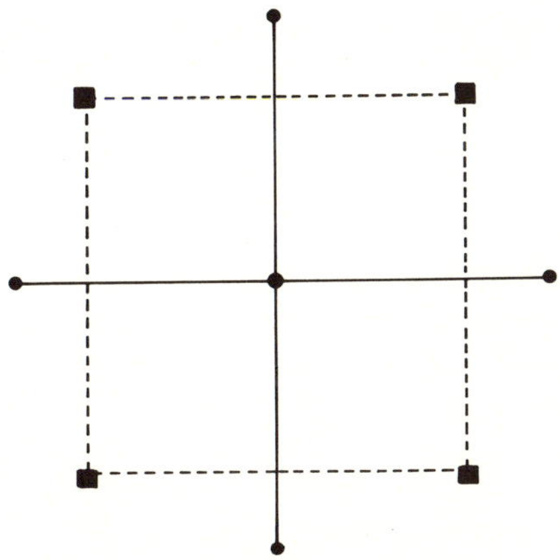

Figure 4.4. Central composite design, a combination of a star (———) and a factorial (----) design

needed. In such cases the three-level factorial design, or the popular central composite design, can be used where the variables can be given (+), (0) and (−) values. The central composite design is illustrated in Figure 4.4 for just two variables. It is called a composite design because it is the centred superimposition of a two-level factorial design and a two-factor star design. It is especially useful since it provides sufficient factor combinations for the full fitting of a second-order polynomial (equation 4.2) and it can be used to approximate almost any smooth, well behaved surface over a limited region.

$$z = a_0 + a_1x + a_2y + a_3x^2 + a_4y^2 + a_5xy + r \tag{4.2}$$

For a central composite design using two variables, nine experiments are required and it soon becomes apparent that a major disadvantage of full factorial designs will be the number of experiments that must be carried out.

A chromatographic example of a three-variable system would be to investigate the effects of pH, ion-pair reagent concentration and mobile phase strength. The particular levels chosen to represent the (+) and (−) levels would be selected on the basis of whatever knowledge is available about the separation. A practical point worth noting is that a (−) level does not imply an *absence* level, but rather a *lower* level. Thus, in the case of an ion-pair reagent, concentrations of 5 and 50 mM might be investigated, not 0 and 50 mM. Taking the two-value, three-variable factorial design shown in Table 4.3, the effect of, for example, variable A can be found in the following way. Looking at experiments 1 and 5, the levels of B and C have been held constant, but the level of A has been changed. The difference between the results provides an estimate of the effect of A when B and C are held constant, both at their higher (+) levels. The difference between the results of experiments 2 and 6 gives another estimate of the effect of A, this time with B constant at its (+) level and C constant at its (−) level. It is easy to see that four estimates of the effect of A are possible from this design. Thus the average effect of A can be found from the eight experiments as being

$$\tfrac{1}{4}[(Y_1 + Y_2 + Y_3 + Y_4) - (Y_5 + Y_6 + Y_7 + Y_8)]$$

where Y_i refers to the result of the experiment i. In similar ways, the effects of the other variables can be found and these are included in Table 4.3.

The next point of interest will be to discover whether any of the variables interact. In other words, does the effect of A depend on the levels of B or C. The difference between experiments 1 and 5 is an estimate of the effect of A at constant B and C, both being at their higher (+) level. The difference between experiments 3 and 7 shows the effect of A with B now at its lower (−) level but again with C at its higher (+) level. If these two differences are compared, this gives an estimate of whether the effects due to A are the same at both levels of B. The greater the difference, the more important are the interactions. Once again it is possible to find four results for the estimate of the interaction between A and B, the net result being given by

$$\tfrac{1}{4}[(Y_1 + Y_2 + Y_7 + Y_8) - (Y_3 + Y_4 + Y_5 + Y_6)]$$

The other two interactions are handled in a similar manner. The final investigation is to determine whether the interactions of A and B are the same at both levels of C and the deduction of the equations for this is left as an exercise for the reader!

The interactions may be difficult to understand from first principles but are well covered in the standard statistical literature.[18–21]. There is a method described by Kamanev which is included as a worked example in ref. 18 which is a very effective way of calculating interactive effects and deducing their significance. This method (also known as Yate's method) is easy to calculate and is well suited to the complete automation of the factorial step.

Having carried out the formal design and experimentation, the most usual next step is to fit the acquired data to a suitable model. This might be a second-order description of the dependence of the capacity factor on the variable levels, e.g. equation 4.2. The model can then be used to generate a response surface of some description, be it simply resolution or a more involved chromatographic response function. Later in this chapter, the simplex lattice design will be discussed: this is a subset of a factorial design. Following experimentation, the results are fitted to a model and the coefficients estimated. A worked example is included and the necessary software is described. Looking ahead to Chapter 5, which discusses the use of 'window diagrams' and minimum alpha plots, it will be seen that factorial designs are sometimes used in this approach as a method for designing the initial experimentation.

4.3.2. Applications

Despite the many uses to which factorial designs have been put in the natural and engineering sciences, the chromatographic literature is surprisingly devoid of applications. Kong *et al.*[22] provided one of the first examples in the optimization of pH and surface-active ion concentration for the separation of a mixture of weak acids, weak bases and zwitterionic compounds. Their interest was in elucidating the combined effects of these variables and factorial design was an obvious formal method of experimentation. To ensure that they obtained as good an estimate as possible, a two-factor, four-level design was used, requiring 16 experiments to be carried out. A worthwhile point made in this paper was that the experiments were carried out in a random order so that effects due to time would not influence the true variable effects. From this work came a simple ion-interaction model, for describing the combined effects of pH and surface-active ion concentration, which could have uses in developing future separations. One potential application of the model might be in its use for the preparation of 'window diagrams' or minimum alpha plots (Chapter 5). This study was also used as an example in a brief discussion of the use of factorial designs in a paper describing multifactor optimization of HPLC conditions.[23]

Model development was also the aim of Otto and Wegscheider,[24] who examined the effects of the mobile phase pH, elution strength and ionic strength on the separation of a selection of ionogenic species with contrasting properties. In order to minimize experimental effort, without compromising the resultant data, they chose a $6 \times 3 \times 2$ factorial design. This means 36 mobile phases were investigated, consisting of all combinations of six pH values, three methanol concentrations and two ionic strengths. In this example, thought was given to the experimental order and the run order was randomized with respect to cost, so that the risk to the column was minimized by avoiding excessively large changes of parameters, and with respect to time. Having acquired the retention information for all these experiments, a semi-empirical model was fitted to describe the retention behaviour of the solutes. The coefficients of this model were used to predict the capacity factors and selectivity factors (alpha values) within the experimental region (pH 2–7, 10 to 30% methanol and ionic strength 0.1–0.2 M). Various methods were then tried in order to locate the optimum chromatographic conditions. The first attempt was by plotting 'window diagrams' and this was followed by minimum alpha plots. Both these methods failed to produce the overall, or global, optimum. The global optimum was found as a result of a computerized grid search of the whole experimental region, the search being the best method of coping with the interdependence of the three variables. The agreement between prediction and subsequent experimental data was excellent.

A central composite design full factorial approach was used by Lindberg *et al.*[25] in the optimization of the separation of a mixture of morphine, codeine, noscapine and papaverine by reversed-phase ion-pair chromatography. In this design, the experiments are positioned at the centre and around the circumference of a circle (see Figure 4.4). The accrued data were fitted to a second-order model which was used to generate a response surface contour plot. This is a useful example for the interested reader to consider in greater detail because the authors discussed the significances of the interaction effects. It was found that, for the molecules under investigation, optimization could be achieved at a fixed separation pH and buffer concentration by manipulating the methanol/water ratio and the concentration of the ion-pair reagent. To estimate the optimum separation conditions the contour plots of capacity factors for individual solutes were overlaid, allowing deduction by inspection. This approach shares much in common with the overlapping resolution mapping method (which will be considered in greater detail in the next section).

A very comprehensive investigation into the factors affecting the separation of the drug sulindac and related compounds by reversed-phase chromatography also exploited the merits of factorial design. Cotton and Down[26] investigated acetonitrile concentration, total organic modifier level (the second modifier being methanol), buffer concentration and separation temperature. Preliminary experiments established the primary variables affecting resolution and the approximate optimum regions for single variables. Next, a four-factor experimental design was used, 25 different experiments being required. Full details of the experimental conditions are given and they provide a useful example of this type of experimental design. A second-order model was used to generate response surfaces. Finally, computer simulation was used to illustrate graphically the effects of the variables on the separation and the column efficiency. The validity of the proposed model was confirmed by experiment and the optimum conditions were located, once again, by a grid search. This search interestingly located two positions of acceptable separation, indicating the problems that can be experienced as a result of there being a number of local optima. The chromatographer can then select the more favourable optimum on the application of a secondary, subjective, criterion such as time, which was not considered in this work.

A modified factorial design investigating three variables at three levels proved successful[27] for the optimization of the reversed-phase separation of inosine and guanosine nucleosides. Twenty-seven experiments were required in order to examine the influence of flow rate, pH and methanol concentration in the mobile phase at constant temperature. A similar design was used for the investigation of methanol concentration, pH and temperature at constant flow rate. For the system under study, the cross-products, that is, the second-order interactions of the variables, were found to be unimportant. The implication here is that the variables are essentially independent and each could be considered in its own right. This will not generally be the case, being dependent on the criterion being employed. However, there will be many situations where, for example, a change in the mobile phase pH over a considerable range will not affect the selectivity of the chromatographic separation.[23]

4.3.3. Software

Factorial designs seem to be so well covered in the standard texts, and presumably in the standard statistical packages that are available on most mainframe computers, that there are no references to programs specifically for HPLC which could be used to form part of an automated system. This seems inconsistent when the potential of the method is considered. For those who enjoy translating programs from one microcomputer to another, software for a Texas Instruments TI 59 programmable calculator to carry out the Yate's analysis is available.[28] Perhaps the dirth of software is indicative of the developments in strategies that have occurred. Thus, factorial design is just a small part in the total optimization approach and has perhaps been overshadowed by other statistical experimental designs and analyses such as the mixture designs which will be considered in the following sections of this chapter. The design itself is not the only point of interest since it is apparent that most workers fit their data to a second-order model. Such fitting programs are well represented in standard statistical packages. A few of the published examples referred to above reference their program sources for model fitting. For example, Otto and Wegscheider used the method of Golub and Pereyre[29] in their model fitting and they carried out the grid search for location of the optimum by a method described by Bevington.[30] (In the next section, the fitting of a model to data from a simplex lattice design will be described and software for this is included as an Appendix.)

4.3.4. Summary

Factorial designs are a means to an end rather than an optimization procedure in their own right. They provide a sound basis for designing experiments and elucidating the relative importance and interactions of the experimental variables.

They can also provide a sound method of investigating conditions surrounding a perceived optimum. Comparing resolution optimization methods, Rafel[31] implied that factorial design should form the first step of method development, to define an empirical description of the system and an area of factor space within which one would explore to locate the true optimum. This is an excellent recommendation but it is just one of many possibilities; the mixture, or lattice, design approach considered next is an alternative.

Advantages of Factorial Designs

1. Efficient experimental design.
2. Indicate the significances and interdependences of individual and groups of variables.
3. Perfectly suited for the fitting of mathematical models.
4. Automation of the data acquisition is relatively easy.

Disadvantages

1. All experiments need to be completed before results can be fully assessed.
2. Full designs with many variables require very many experiments.
3. Latent effects may confuse the results if the experimental order is not carefully randomized.
4. Important information may be missed if experimental points are either too far apart or too close together.

4.4. MIXTURE DESIGN STATISTICAL TECHNIQUES AND OVERLAPPING RESOLUTION MAPPING

4.4.1. Theory

One of the problems associated with the use of factorial designs can be the number of experiments that are required to provide data for a suitable model. Fortunately, it is possible to conduct fewer experiments than are required for a full factorial design by using a *mixture design* statistical approach. An initial factorial experiment may still be required for the identification of key variables. With or without some preliminary experimentation, problems still arise in the selection of a small number of solvents with which to commence method development. For reversed-phase chromatography the problems are minimized as three modifiers are likely to satisfy most requirements. Since reversed-phase chromatography remains the firm favourite for most workers, the use of mixture designs, together with the overlapping resolution mapping approach (ORM), will be considered by reference to reversed-phase applications.

The ultimate goal in the ORM approach to separation optimization is to permit the selection of a mobile phase composition that provides the best resolution between the worst separated pair within a desired capacity factor range. This aim is achieved by manipulation of the selectivity of the mobile phase while maintaining an approximately constant solvent strength. From the selectivity triangle (Figure 4.2), the solvents that will most commonly be used as modifiers for reversed-phase chromatography will be methanol, acetonitrile and tetrahydrofuran. Water will be used as the carrier to maintain the equivalent solvent strength. Working with three organic modifiers, the three binary solvent compositions (water + modifier) will define the three vertices of an optimization triangle. In the mixture design, seven to ten experiments are sufficient to evaluate the entire selectivity region that is defined by this triangle. Each of the experiments is carried out with a mobile phase intended to produce a similar k' range for all the components of the mixture, although (hopefully) specific retention orders and selectivities will change. Figure 4.5 shows the optimization triangle and the solvent blend ratios that would be used for a ten-experiment design. This particular experimental design is known as a *simplex lattice design*.[32] Early applications of the ORM approach used a design which involved ten experiments spaced throughout the optimization triangle as shown in Figure 4.5, but subsequently seven experiments have been found to suffice, particularly where solute identification is included. The omission of the final three experiments simply prevents the investigation of the fit of the experimental data to the derived model.

The first practical position to be established is an estimate of the required solvent strength. There are several ways in which the solvent strength necessary to produce an isocratic separation within a given time can be estimated. One of the better ways is to run an initial gradient from which the desired solvent strength can be deduced (see Section 3.3 in Chapter 3). It is possible, with suitable chromatographic hardware, to carry out this first step and the subsequent

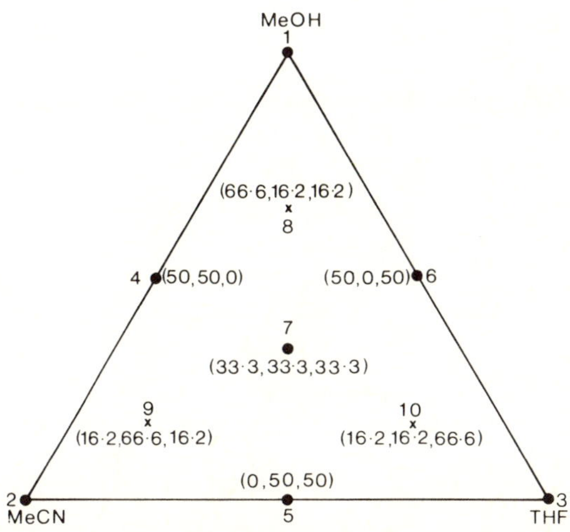

Figure 4.5. Simplex lattice design for reversed-phase chromatographic optimization showing relative proportions of each solvent to be used

calculations entirely automatically and this step forms the first 'scouting' part of the optimization.[33]

Referring to the six-component test mixture, the aim will be to optimize the separation in a total analysis time of approximately 3 minutes ($k' = 5$). An isocratic mobile phase of methanol/water (60:40) achieves elution of all the peaks within this time constraint.

The next step in the procedure is to define the other corners of the triangle by calculating solvent compositions of the other two modifiers (acetonitrile and tetrahydrofuran) with water, which should have the same total solvent strength. Theory predicts that mobile phases of the same solvent strength will produce approximately equal capacity factor ranges. There are several ways in which equivalent solvent strengths can be calculated. In summary, the two most successful approaches are likely to be those described by Schoenmakers *et al.*[34] (equations 4.3 and 4.4) and Snyder *et al.*[35]

$$\phi_{\text{MeCN}} = 0.32\phi^2_{\text{methanol}} + 0.57\phi_{\text{methanol}} \tag{4.3}$$

$$\phi_{\text{THF}} = 0.66\phi_{\text{methanol}} \tag{4.4}$$

where ϕ is the volume fraction of the solvent.

The Snyder approach is slightly more complex but arguably more versatile. Here a parameter (ST) is used to describe the total solvent strength of the mobile phase, that strength consisting of the sum of the weightings of the individual components multiplied by their proportions. To work out a new composition, the value of ST must remain constant. The equation is

$$ST = s_a\phi_a + s_b\phi_b + \ . \ . \ . \tag{4.5}$$

where s is the solvent strength weighting factor, being 2.6 for methanol, 3.2 for acetonitrile, 4.5 for tetrahydrofuran and 0 for water.[1,35] To show how equation 4.5 works in practice, consider the mobile phase methanol/water (60:40). The total solvent strength comes from

$$ST = s_{\text{methanol}}\phi_{\text{methanol}} + s_{\text{water}}\phi_{\text{water}}$$

$$ST = (2.6 \times 0.6) + (0 \times 0.4) = 1.56$$

To calculate the required proportion of tetrahydrofuran:

$$1.56 = s_{\text{THF}}\phi_{\text{THF}} + s_{\text{water}}\phi_{\text{water}}$$

i.e.

$$\phi_{\text{THF}} = 1.56/4.5 = 0.347 \text{ or } 34.7\%$$

Continuing with the 60% methanol example, equations 4.3 and 4.4 would predict 45.7% and 39.6% for acetonitrile and tetrahydrofuran, respectively, whereas from equation 4.5 the values would be 48.7% and 34.7%. For two

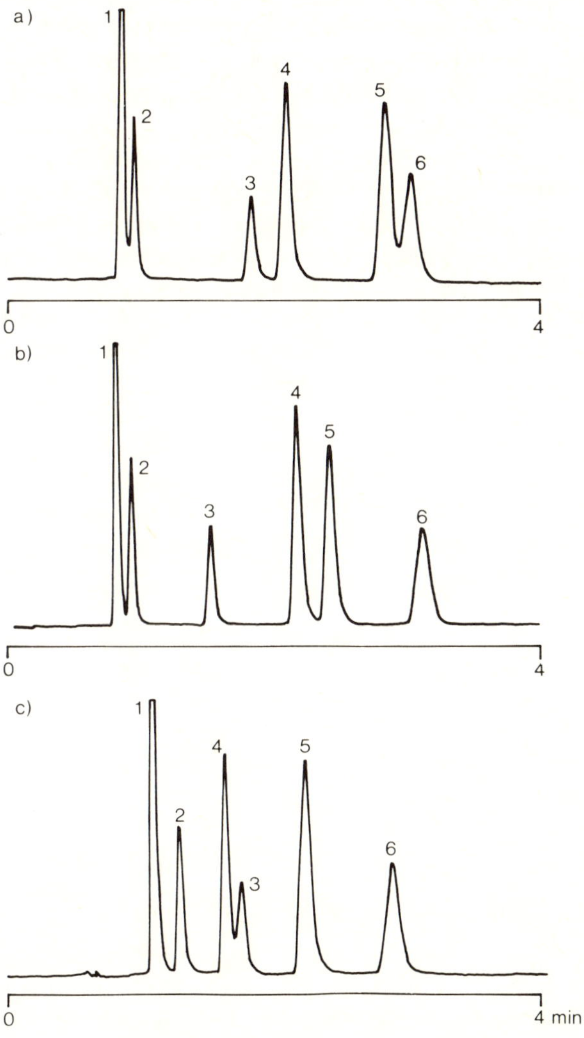

Figure 4.6. Chromatograms of the six-component mixture obtained with iso-elutropic mobile phases: (a) 60% methanol; (b) 48% acetonitrile; (c) 42% tetrahydrofuran

very contrasting approaches the agreement is adequate when it is remembered that it is the *average* solute which is being considered—in any case, the hope is that individual solutes will not behave in an average manner. Figures 4.6a–c show the chromatograms of the test mixture run with isocratic mobile phases containing 60% methanol, 48% acetonitrile and 42% tetrahydrofuran, respectively. All three chromatograms elute the last component within the desired time constraint but it was necessary to increase the proportion of tetrahydrofuran considerably to achieve the desired analysis time. It is also obvious that all solutes do not behave equally with the different modifiers. This is indeed fortunate, for it is the very essence of the selectivity effects and differences that are to be exploited.

For normal-phase chromatography, the situation is less straightforward, particularly if solvent localization effects are to be taken into account. Equation 4.5 will be satisfactory for most situations, except where solvents with large localization effects are being used. For those interested in calculating solvent strengths in ternary and quaternary solvent systems, especially for the cases with strong polar solvents which can localize, then a paper by Glajch and Snyder[36] and a review on mobile phase effects in liquid–solid chromatography[12] are both essential reading.

Having established the three extremes of the experimental space, any other experiments carried out within this region, that is, using blends of the three binary mixes, will be expected to give similar total retention times. This is because the total solvent strength, or polarity, will be approximately constant—this follows from a consideration of equation 4.5. In practice, it will be found that small adjustments to the binary solvent composition will probably be required in order to account for specific solute/stationary phase properties, and this has already been demonstrated in the need to adjust the tetrahydrofuran content for the example above.

The next three experiments to be conducted represent the three mid-points of the sides of the triangle (Figure 4.5). Where a seven-point simplex design is being used, a final, seventh, experiment is then carried out at the centre of the triangle. These seven definitive experiments can all be automatically undertaken by a computer-controlled instrument; even a relatively modest apparatus could be programmed to carry out the chromatography, although no subsequent data analysis would be possible.

It is the data analysis and chromatogram evaluation that are the most difficult aspects of the ORM approach. Having completed the search for seven experimental points within the experimental domain, the retention behaviour of all the peaks needs to be described. If standards are available for all the components of the mixture then they should be injected under the experimental conditions corresponding to the seven points. If standards are not available, then some manual method of peak identification is likely to be required. Some authors have described attempts to use peak areas[37] to identify and monitor peak movements during optimization runs, but this technique is far from reliable. Other methods worth considering include the use of multi-channel detection systems which give information on the identity of the peaks. An obvious choice here is the diode-array UV spectrophotometer, but one must bear in mind the dramatic changes that can occur in absorption spectra as a result of solvent changes.

A simplified procedure, which does not actually necessitate consideration of the resolution mapping of the system, uses a chromatographic response function. In the original work describing the mixture design approach, Glajch *et al.*[3]

began by using the chromatographic optimization function (COF) described by equation 4.6:

$$\mathrm{COF} = \Sigma A_i \ln (R_i/R_{id}) + B(t_M - t_L) \qquad (4.6)$$

where R_i is the resolution of the ith pair, R_{id} is a desired resolution for that pair, t_M and t_L are the retention times desired for the last-eluted peak and its actual retention time, respectively, and A and B are arbitrary weighting factors. In use, A was set to unity and B was set to zero, making the COF very much easier to calculate and manipulate. This equation, or something similar selected after reading Chapter 2, can form the basis of a very helpful automated system with suitable experimental design.

Glajch *et al.*'s procedure calculates the COF for each of the seven chromatograms described by the simplex design and then fits the best second-order polynomial equation to the four variables, that is, the three solvent compositions and the dependent COF. This method permits the optimization of a separation without requiring that the peaks be identified and it is well suited to an automated system. Indeed, the calculations are almost trivial and can be handled by even the most modest programmable calculator or computer. For those not familiar with the procedure for fitting a cubic model to experimental data from a design such as the simplex, the following example describes the essential steps.

1. The three vertices of the solvent selectivity triangle corresponding to equal solvent strength are defined. In this example these mobile phases are 60% methanol, 48% acetonitrile and 42% tetrahydrofuran—these will be referred to as solvents A, B, and C, respectively.
2. Seven experiments are conducted according to the simplex lattice design (see Figure 4.5) and the chromatograms are evaluated according to a suitable criterion, e.g. the COF of Glajch *et al.*
3. The results are used to define the chromatographic response surface by fitting a cubic model.
4. The optimum composition(s) is (are) found either by a stepwise search or by a graphical procedure.

Steps 1 and 2 are summarized in Table 4.4, which describes the seven experimental conditions and the COF values obtained for these experiments.

TABLE 4.4. Mixture design optimization of six-component mixture

Experiment No.	A(%)	B(%)	C(%)	COF
1	100	0	0	1.34
2	0	100	0	2.78
3	0	0	100	1.63
4	50	50	0	0.64
5	0	50	50	−0.35
6	50	0	50	1.85
7	33	33	33	1.37

```
PLOT OF MIXTURE RESPONSE SURFACE FOR 6 COMPONENT TEST MIX
================================================================

COEFFICIENTS OF POLYNOMIAL Y = AX+BY+CZ+DXY+EXZ+FYZ+GXYZ ARE...

COEFFICIENT  1   1.34
COEFFICIENT  2   2.78
COEFFICIENT  3   1.63
COEFFICIENT  4  -5.68
COEFFICIENT  5  -7.34
COEFFICIENT  6  -1.42
COEFFICIENT  7  28.56

RESPONSE IS COF of Glajch

PLOT LEVEL UP TO  0       = .
PLOT LEVEL UP TO  1       = -
PLOT LEVEL UP TO  2       = +
PLOT LEVEL UP TO  2.5     = *
PLOT OVER LEVEL   2.5     = #

MAXIMUM RESPONSE IS  2.78  AT  0 %: 100 %:  0 %

                                          0
                                        5 +  100
                                     10 + +  95
                                   15 - + +  90
                                 20 - + + +  85
                               25 - - + + +  80
                             30 - - + + + +  75
VARIABLE: 60% MeOH         35 - - - + + + +  70
                         40 . - - + + + + +  65
                       45 . - - - + + + + +  60
                     50 . - - - + + + + + +  55
                   55 . - - - + + + + + + +  50
                 60 . - - - - + + + + + + +  45
               65 . - - - - + + + + + + + +  40
             70 . - - - - + + + + + + + * *  35
           75 . - - - - + + + + + + + + * *  30
         80 - - - - - - + + + + + + + * * *  25
       85 - - - - - - + + + + + + + + * * *  20
     90 - - - - - - - + + + + + + + + * * *  15
   95 - - - - - - - - - + + + + + + + * * #  10
100 + - - - - - - - - - - + + + + + + * * #  5
   + + - - - - - - - - - - - - + + + + * * #  0
   0                          48% MeCN --->     100
```

Figure 4.7. Computer plot of response surface (calculated by the program in Appendix IV) for simplex lattice design using combinations of three mobile phases described in Figure 4.6

In using the COF of Glajch *et al.*, the values of parameters A and B were set to unity and zero, respectively, and the minimum desired resolution was also set at unity.

The response surface is described by the following equation:

$$Y = a_1X_1 + a_2X_2 + a_3X_3 + a_{12}X_1X_2 + a_{13}X_1X_3 + a_{23}X_2X_3 + a_{123}X_1X_2X_3 \qquad (4.7)$$

where Y is the COF value and the a coefficients are found experimentally. The calculation of the a coefficients is shown in the following equations:

$$a_1 = Y_A \qquad (4.8)$$
$$a_2 = Y_B \qquad (4.9)$$
$$a_3 = Y_C \qquad (4.10)$$
$$a_{12} = 4Y_{AB} - 2(Y_A + Y_B) \qquad (4.11)$$
$$a_{13} = 4Y_{AC} - 2(Y_A + Y_C) \qquad (4.12)$$
$$a_{23} = 4Y_{BC} - 2(Y_B + Y_C) \qquad (4.13)$$
$$a_{123} = 27Y_{ABC} - 12(Y_{AB} + Y_{AC} + Y_{BC}) + 3(Y_A + Y_B + Y_C) \qquad (4.14)$$

By substituting the values of a in equation 4.7, the value of Y (i.e. the COF) at any composition can be found, assuming that equation 4.7 is an accurate model of the response surface. Figure 4.7 is the output from a computer program which does just this (see Appendix IV); the maximum values of the COF are to be found at high acetonitrile proportions.

This use of the mixture simplex lattice design in conjunction with the COF is well suited to automated optimization because of its great simplicity. Unfortunately, it suffers from the limitation common to most chromatographic optimization functions—it cannot adequately cope with peak crossovers. This restriction was recognized by Glajch *et al.*, who proposed a new data analysis method based on preparing resolution contour plots.

The new chromatographic analysis technique relies on the measurement and comparison of the resolution between every pair of peaks in the chromatograms obtained for the (usually seven) solvent compositions investigated. A resolution contour plot is prepared for each pair of compounds depicting the resolution for that pair in all compositions. A desired minimum resolution (e.g. 1.5) between peak pairs is then specified and this enables areas on the contour plot to be eliminated as being of no further interest since they do not meet the desired criterion. Figure 4.8 shows composition regions (shaded) for each of the adjacent peak pairs that fail to meet the resolution criterion. By superimposing all the specific resolution maps of the various peak pairs, areas defining particular solvent compositions can be identified in which the desired separation can be achieved for all solutes in the mixture. Mathematically defined, the ORM method utilizes the intersection of Venn diagrams describing acceptable resolutions for the compounds of interest.

For the example of the six-solute test mixture, five resolution plots are generated (Figure 4.8). The generation of these plots obviously required that

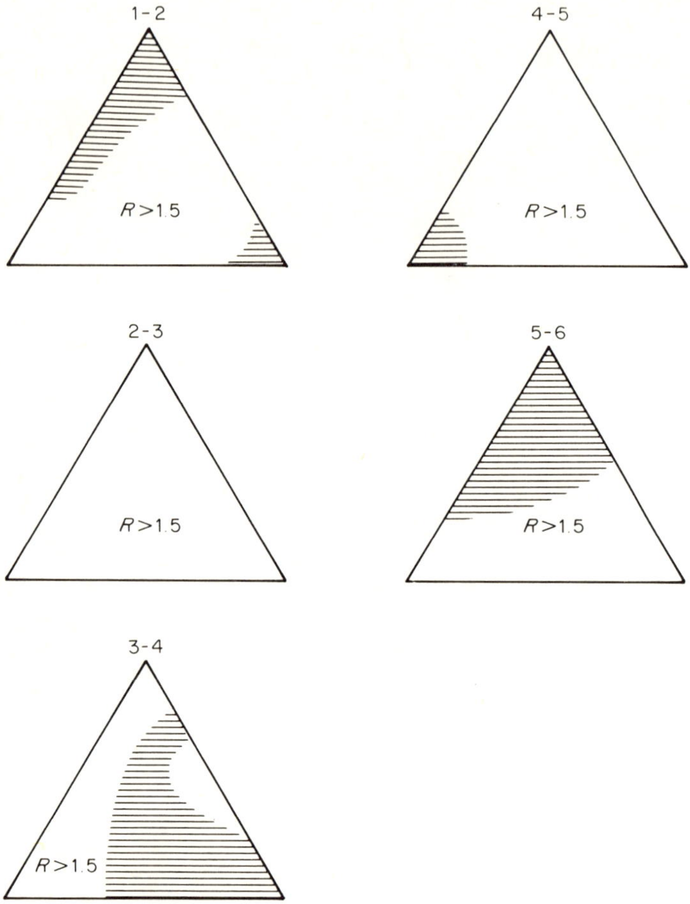

Figure 4.8. Individual resolution maps for the five pairs of solutes in the test mixture

every peak was identified in each chromatogram. This is straightforward in this particular example but may well require that standards of every solute be chromatographed for more complex examples. The unshaded areas represent resolutions greater than 1.5 and it can be seen that just one peak pair (2–3) always exceeds this minimum criterion. Only the peak pairs that do not meet this criterion need be considered further and Figure 4.9 shows the ORM for these peaks. The unshaded regions (marked A and B) are the areas within which separations should give resolutions between all peak pairs in excess of 1.5. It is interesting to compare Figures 4.9 and 4.7 to see the contrasts in the results. Whilst Figure 4.7 directs attention to a relevant area which indeed will give an

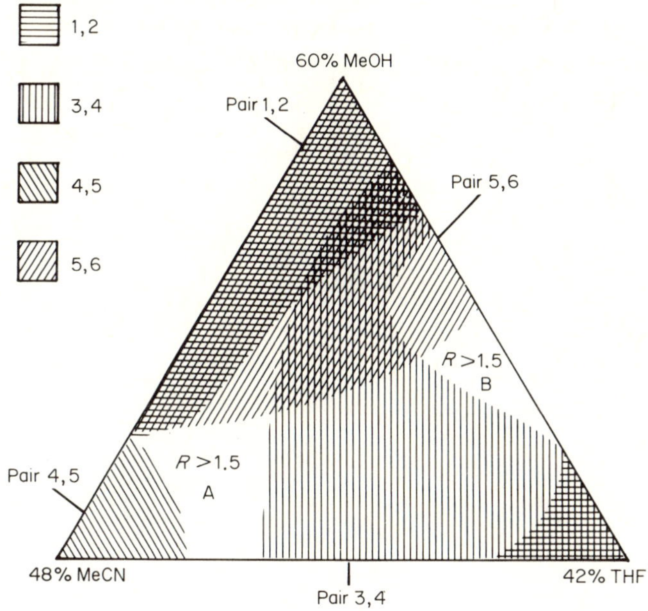

Figure 4.9. Overlapping resolution map (ORM) for the six-component test mixture: data from Figure 4.8

adequate separation, no indication is given of the very poor regions of composition where peak crossover is occurring.

It should be noted that there is additional information available from Figure 4.9. If one peak is removed from the consideration (e.g. peak 3), then the range of suitable compositions is greatly increased. A disadvantage of the procedure (removed in the extension of the method to three-dimensional window diagrams; Chapter 5) is that there is no indication as to precisely where one should choose to work, since the ORM does not define a unique solution. This is evident by reference to Figure 4.10, which shows chromatograms from both regions A and B. It is obvious that both chromatograms represent satisfactory solutions to the immediate problem but they differ in the order of elution of the peaks. It would be up to the chromatographer to decide which was the more acceptable separation based on a secondary consideration such as mobile phase costs or perhaps even toxicity or odour of the solvents. One further aspect that is not fully considered is the optimization of time. The ORM method intentionally works with a constant, pre-specified, time constraint. Time optimization is given more consideration in other refinements of the mixture design approach.

The whole ORM procedure can be conducted completely automatically by a suitable computer-controlled chromatograph and indeed formed the basis of the SENTINEL system which was developed by the Du Pont Company in the

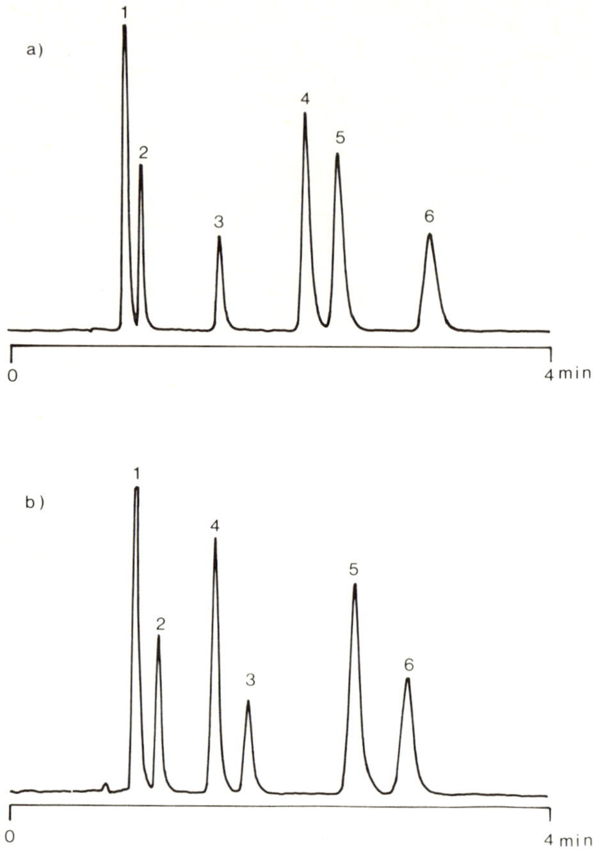

Figure 4.10. Separations obtained under indicated optimum compositions from Figure 4.19: (a) from area A and (b) from area B

early 1980s.[33] The major problems with the fully automated approach lie in the need to identify components, if the procedure is to be taken to its ultimate conclusion, and the high probability that all solvent areas may be excluded for complex samples. The procedure itself is, however, fairly general and can be applied to normal-phase chromatography[38] just as well as to reversed-phase separations, provided that the solvent strength can be calculated. For normal-phase chromatography, gradient 'scouting' is more difficult to carry out although a method was described by de Smet *et al.*[39] (see Section 3.3 of Chapter 3). The first step may have to be one of intuitive exploration in order to establish the required solvent strength. Perhaps too, in normal-phase chromatography, the emphasis for the choice of solvents to describe the selectivity triangle should concentrate on the localization effects as described in the section on the classification of solvents.

The ORM method is not restricted to use in isocratic separations. Kirkland

and Glajch[40] showed that gradient optimization can be carried out using the ORM method. There is a problem here in trying to define just what constitutes a gradient separation. Glajch and Kirkland[41] sought to establish a convenient classification of multi-solvent mobile phase systems for liquid chromatography. The general description is intended to enhance a more systematic investigation of solvent strength and selectivity effects and to facilitate the efforts in HPLC optimization. The new classification recognizes that earlier definitions of gradient elution referred to changes in solvent composition which also led to changes in solvent strength. Such solvent strength changes are not inevitable in dealing with ternary and more complex mobile phases. It is possible to change the mobile phase composition and keep the solvent strength constant, changing only the selectivity. The four categories of solvent systems that have been proposed are summarized in Table 4.5. The first two maintain a constant solvent strength while the last two consider variations in solvent strength.

TABLE 4.5. Classification of solvent systems[41]

Type	Name	Condition during separation
A	Simple isocratic (SI)	Solvent strength, selectivity and composition are constant
B	Isocratic multi-solvent programming (IMP)	Solvent strength is constant; solvent selectivity and composition change
C	Isoselective multi-solvent gradient elution (IMGE)	Solvent strength and composition change; selectivity remains constant because the ratio of the modifiers is constant
D	Selective multi-solvent gradient elution (SMGE)	Solvent strength, composition and selectivity all change

These four mobile phase systems are illustrated for reversed-phase chromatography in Figure 4.11. It is immediately obvious that a solvent selectivity prism is being considered for gradient elution, rather than the planar triangle which has been used for isocratic separations. It is probable that most of the multi-solvent 'gradient' separations which have been described fall into the IMGE category. Thus, during the separation, for reversed-phase chromatography the proportion of water will have been reduced and the relative proportions of the organic modifiers will have been increased in some (arbitrary) way. The systematic optimization of the IMGE separations can be tackled in a manner entirely analogous to the ORM method used for isocratic separations. The selectivity triangle is now replaced with a selectivity prism[40] and seven experiments are conducted according to the design indicated in Figure 4.12. A quadratic model is used to describe the response surface, the values of the coefficients being deduced from the data acquired from the seven experiments. As with the optimization of isocratic separations, this simple simplex design produces sufficient data to define the resolution surface of the selectivity prism. The computer software needed is the same as that used in the isocratic case,

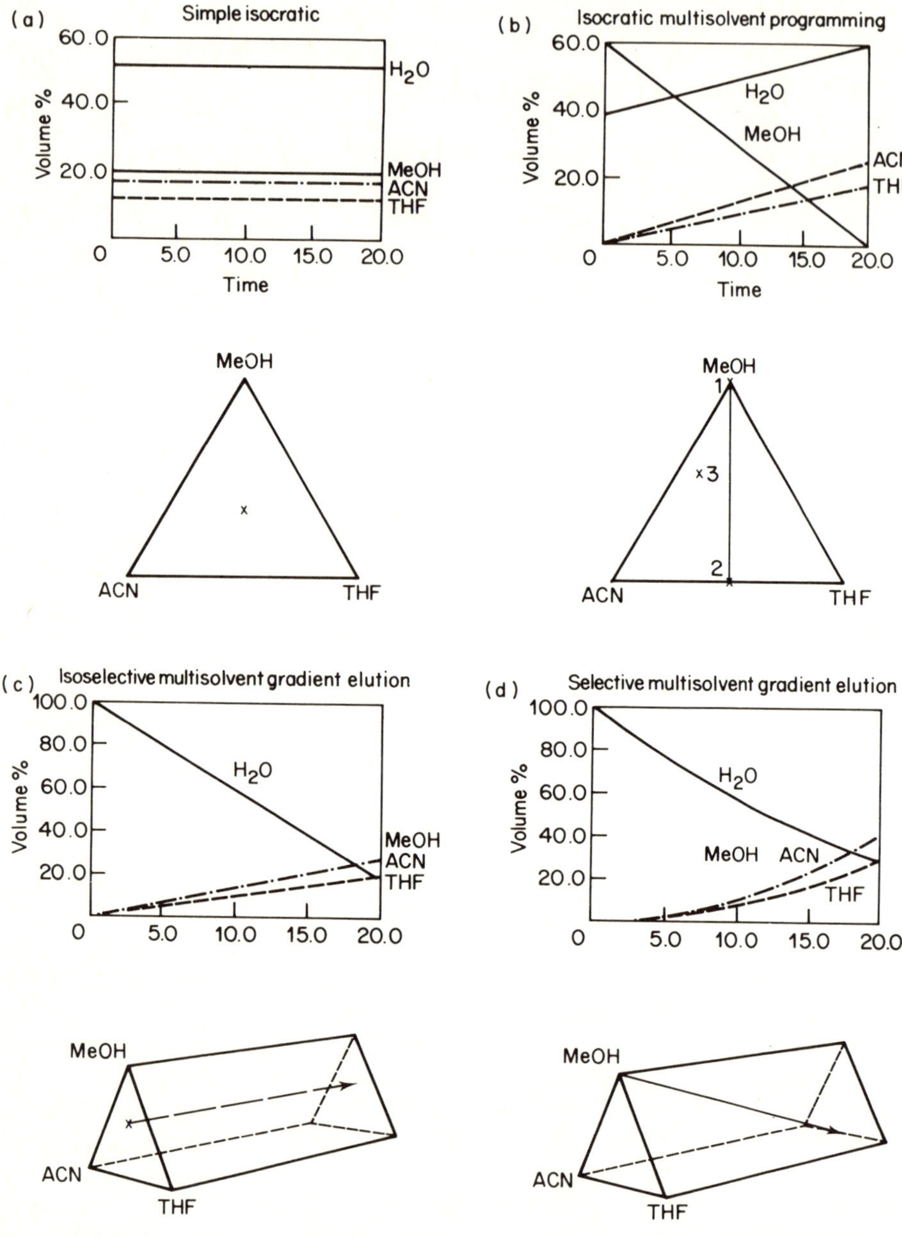

Figure 4.11. Examples of the four classifications of mobile phase systems for reversed-phase HPLC with typical composition/time profiles: (a) simple isocratic; (b) isocratic multi-solvent programming; (c) isoselective multi-solvent gradient elution; (d) selective multi-solvent gradient elution. *Reprinted with permission from Ref. 41. Copyright (1982) American Chemical Society*

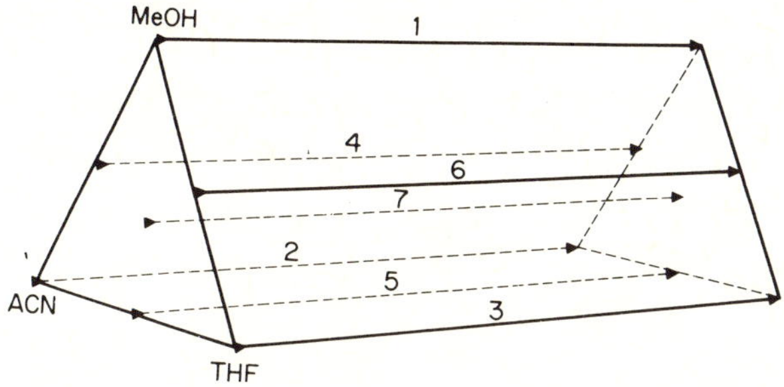

Figure 4.12. Experimental design for seven gradient runs to obtain basic data for separation optimization. *Reproduced from ref. 40 by permission of Elsevier Science Publishers BV*

although it is necessary to use the constant increase in solvent strength, dS/dt, rather than the solvent strength, S, required in isocratic separations. The optimum conditions are located in an exactly equivalent manner. The resolution contour plots for all peak pairs are overlaid and that area providing the greatest minimum resolution represents the optimum operating conditions.

The most general type of solvent system for gradient elution will be selective multi-solvent gradient elution, SMGE. In this case everything is altered! Solvent strength, selectivity and composition are simultaneously varied during the separation and this must provide the greatest potential for the most difficult separations. Fully automated optimization of such separations will be difficult, unfortunately, because of the infinite variety of profiles that can be examined. It is not necessary to be limited to linear changes or even continuous functions. Step changes in composition can be of great use. Much of the optimization in SMGE separations will have to be carried out visually. It will be, however, extremely advantageous, if not essential, to carry out the formal design described in the IMGE case so as to define the nature of the problem.

Finally, the ORM method can be used in the selection of the stationary phase as well as the mobile phase composition.[42] What is done now is to superimpose mobile phase selectivity triangles on top of a stationary phase selectivity triangle (Figure 4.13). It has been argued[17] that the stationary phase, like the mobile phase, involves related variables since the sum of the components of the stationary phase must be 100%. One should be aware that the selectivity differences likely to be seen as a result of stationary phase changes in bonded-phase chromatography will probably be small in relation to the selectivity changes that will arise as a consequence of mobile phase effects. Nevertheless, with modern computer-controlled instrumentation, the effort required to carry out the 21 experiments now required (a seven-experiment simplex design for each of three stationary phases) is small.

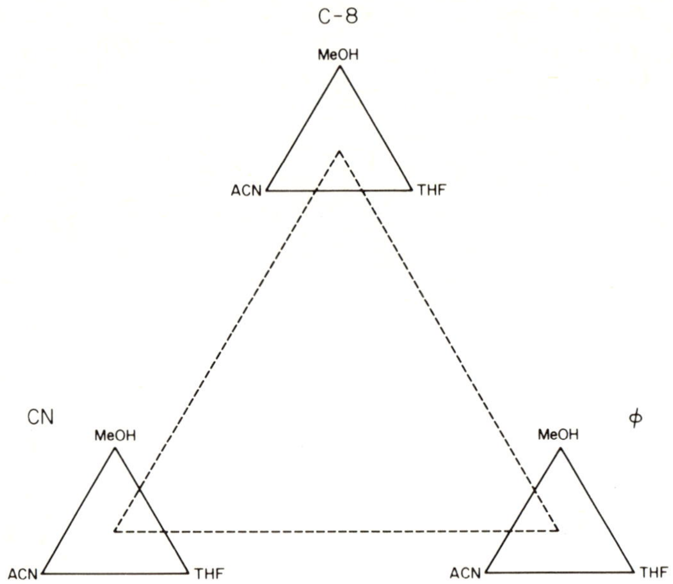

Figure 4.13. Experimental design for combined optimization of stationary phase and mobile phase in reversed-phase HPLC. *Reproduced from ref. 42 by permission of Elsevier Science Publishers BV*

4.4.2. Applications

The description of the optimization of a normal-phase separation of a steroid mixture by Antle[38] is highly recommended reading, as each step in the process is well described and documented. The optimum region as defined by the map is extremely small but successfully located by the method.

Landers and Olsen[43] used the ORM method to optimize the separation of the geometric isomers of retinol using normal-phase chromatography. Dichloromethane, diisopropyl ether and chloroform were chosen as the polar modifiers in accord with the selectivity triangle (but not the localization triangle). The resolution values for each isomer (there being four components in total) were fitted to a quadratic model of resolution as a function of solvent composition. From this model the resolution contour plots were produced and the final optimum conditions were determined by manually overlaying the various plots.

Glajch *et al.*[10] used the solvent localization triangle in optimizing the normal-phase separation of 13 substituted naphthalenes. Optimization in relation to time was also discussed with the two approaches of either increasing the flow for the whole separation, or using flow programming, being shown.

A good example of the method in action in reversed-phase chromatography can be found in the original paper of Glajch *et al.*,[3] which has been described in detail already. A ten-experiment design was used here to establish the semi-empirical model. The paper also described an analysis of another literature separation, but it is disappointing to discover that the chromatogram corresponding to the optimum conditions is computer simulated.

A more challenging task was attempted by Lurie *et al.*,[44] who used ORM to optimize the reversed-phase separation of 26 fentanyl homologues and analogues. Although the total separation of all 26 compounds was not achieved, conditions of maximum selectivity were established and the discriminating power of the separation was enhanced by employing absorbance ratioing, using two detectors in series. The factors affecting the selectivity of this separation were further studied[45] using the solvent mixtures employed in the ORM scheme. The selectivities obtained from reversed-phase separations on conventional silica-based ODS columns were compared with those for a polymer-based PRP-1 column. The study concluded that, for these two reversed-phase systems, there was little difference in the selectivities produced by the two columns.

The separation of the three constituents of Cardiotensin (bemetizide, triamterine and bupranolol) was optimized using a simplex lattice design

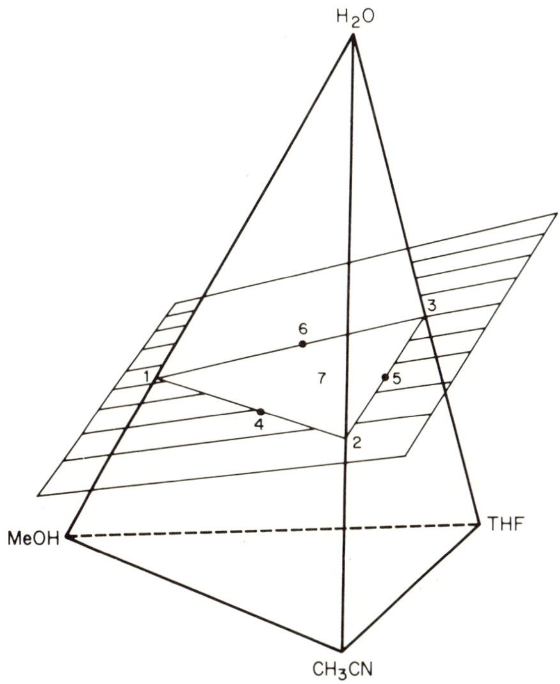

Figure 4.14. Three-dimensional representation of the range of mobile phase compositions available with four solvents. The conditions normally searched by combination of iso-elutropic mobile phases corresponds to the plane intersecting the tetrahedron. *Reproduced from ref. 47 by permission of Elsevier Science Publishers BV*

experimental procedure.[46] Optimum conditions were deduced by inspection rather than by using the ORM method.

The solvent optimization of the reversed-phase separation of polar adrenal steroids was described by D'Agostino *et al.*,[47] who used an ORM method. This work is particularly interesting as it considers a complex mixture and also compares the ORM concept with the use of a modified form of the COF (equation 4.6). In this case the authors chose to investigate the value of using weighting factors in the COF and were able to predict an excellent separation. However, they pointed out a very interesting feature of the basic ORM method, namely that the original method only explores a plane of composition (Figure 4.14) and, in doing so, cannot guarantee that the optimum separation actually lies on this plane. It is well known that the addition of water to organic solvents changes their bulk properties, resulting in effects that might not have been expected. For example, considerable changes in temperature occur on mixing methanol or acetonitrile with water and there are variations in the viscosity of mobile phases with differing methanol/water ratios. A suggestion for

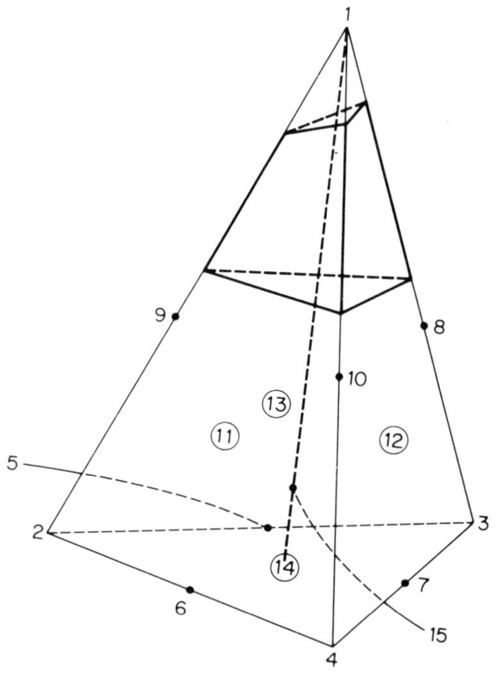

Figure 4.15. Statistical design for searching the complete range of mobile phases afforded by four solvents. The numbers indicate individual mobile phase compositions at which retention data for all solutes must be obtained. In isocratic, reversed-phase chromatography it may well be possible to eliminate some water-rich and modifier-rich conditions; in this case a twelve-point design, within the region represented by the truncated pyramid, will be sufficient. *Reproduced from ref. 47 by permission of Elsevier Science Publishers BV*

overcoming the planar limitation is not to restrict the optimization to pre-selected solvent strengths. It therefore becomes necessary to search the full quaternary system that is represented in three-dimensional space by the tetrahedron shown as Figure 4.15. The simplex design is then expanded to 15 experiments to cover the whole domain. It should not be essential to carry out the whole of this design since previous chromatographic experience may allow one to rule out extreme compositions of water-rich or water-lean eluents. If this is the case, then a truncated pyramid can be searched and this can be done with just twelve experiments.

The extension of the ORM method to gradient elution was described by Kirkland and Glajch,[40] who found it advantageous to classify just what is meant by 'gradient' elution. They found the ORM method easy to apply to IMGE solvent systems, that is, where the selectivity is held essentially constant

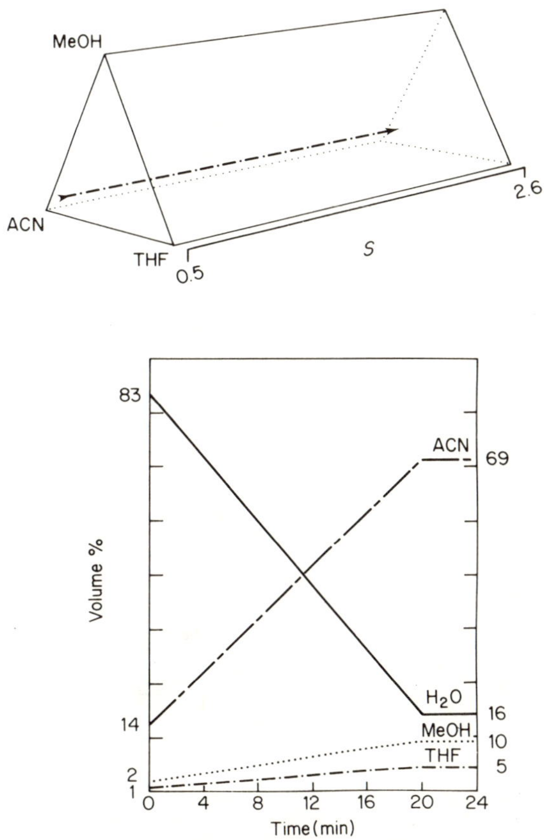

Figure 4.16. Schematic representation of the solvent programme for the optimum IMGE solvent system. *Reproduced from ref. 40 by permission of Elsevier Science Publishers BV*

during linear changes in solvent strength. Just seven experiments were required to define the resolution surface and allow the resolution contour plots to be prepared. The eighth experiment gave a good separation of the 14 components with wide polarity differences that were being investigated. The solvent programme is illustrated in Figure 4.16, with the optimized separation shown as Figure 4.17. Kirkland and Glajch[40] extended this approach further to the optimization of selective multi-solvent gradients but here it was necessary to carry out the selection of optimum conditions visually. The data accumulated from the IMGE experimentation indicated that further improvement in the separation should be possible, particularly if step changes in solvent composition were employed. Three step changes were selected, depicted in Figure 4.18. The selection of this particular profile surely reflects the influence of extensive chromatographic expertise but does indicate some of the dramatic effects that can be exploited in gradient elution. Inspection of the optimized separation (Figure 4.19) shows that all peaks are now well separated but with some substantial changes in elution order. Note, however, a possible disadvantage resulting from the dramatic and sudden changes in solvent composition, i.e. the base-line perturbations that have occurred in the middle of the run.

Glajch and co-workers described the application of the ORM approach to the simultaneous optimization of mobile and stationary phases.[17,42] They examined octylsilane (C_8), phenylethyl- and cyano-bonded columns in the

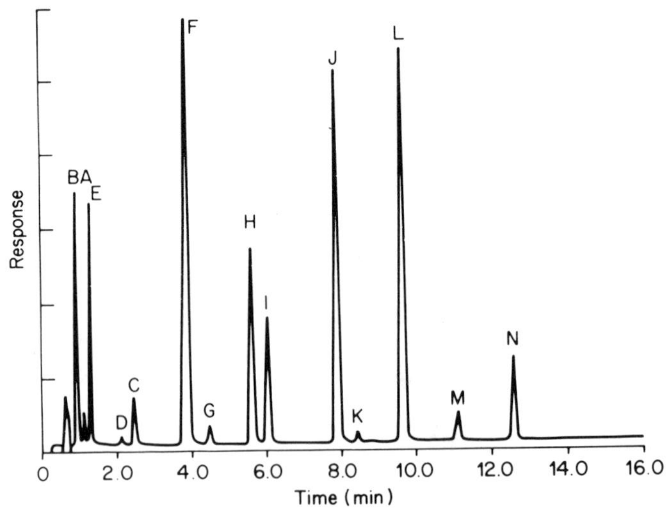

Figure 4.17. Chromatogram for optimum IMGE separation. Solute identities (A) resorcinol; (B) theophylline; (C) phenol; (D) benzyl alcohol; (E) caffeine; (F) methyl paraben; (G) benzonitrile; (H) nitrobenzene; (I) cortisone; (J) propyl paraben; (K) Ramrod; (L) butyl paraben; (M) chloroisopropyl *N*-(3-chlorophenyl) carbamate; (N) progesterone. *Reproduced from ref. 40 by permission of Elsevier Science Publishers BV*

optimization of the separation of a 20-component mixture of phenylthiohydantoin (PTH) amino acids.

Despite the classification of variables such that pH was considered to be a discrete, rather than related, variable, ORM has been used in the optimization of the ion-pair separation of bases.[48] In this case the selectivity triangle was formed from pH 2.5 buffer, pH 7.5 buffer and aqueous n-hexanesulphonic acid solution, each being adjusted with methanol to give approximately constant elution times for the last-eluted peak. Samples successfully separated using this approach include tricyclic antidepressants, water-soluble vitamins and constituents of

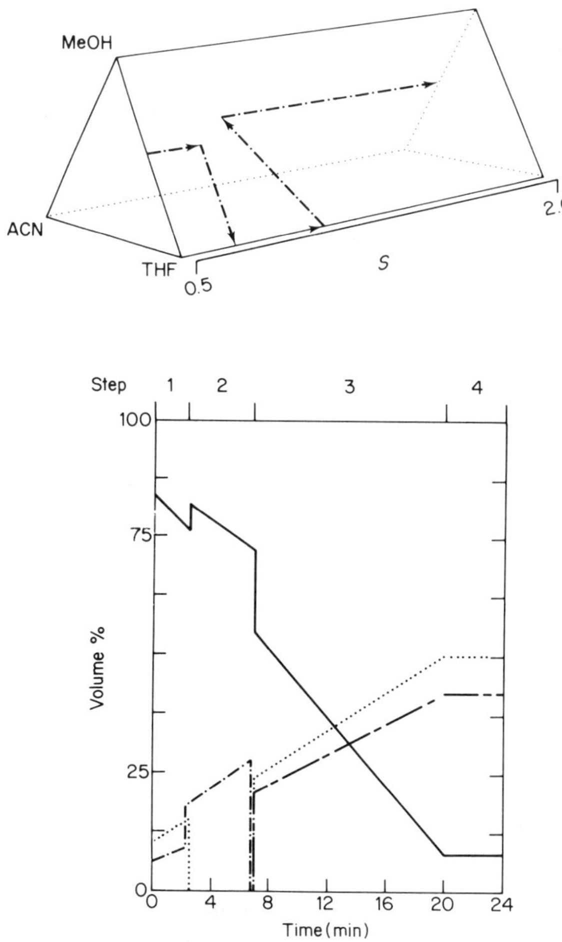

Figure 4.18. Schematic representation of solvent programme for step-selectivity gradient solvent system employing water (———), methanol (· · · · · ·), acetonitrile (—— — ——) and tetrahydrofuran (— · — · —). *Reproduced from ref. 40 by permission of Elsevier Science Publishers BV*

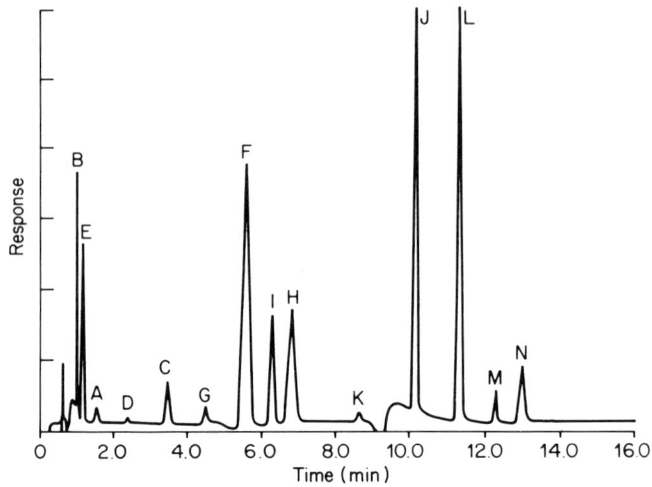

Figure 4.19. Chromatogram obtained with step-selectivity gradient shown in Figure 4.18; note baseline perturbation after peak K. Solute identities as in Figure 4.17. *Reproduced from ref. 40 by permission of Elsevier Science Publishers BV*

common cold remedies. This paper is of particular interest to all involved in the application of ion-pair separations since it is concluded '. . . *in principle, a single pairing ion can provide all the selectivity available from a range of different pairing ions, providing that its concentration can be varied to change the fractional coverage of the stationary phase by the reagent over a wide range.*' Further, the authors conclude that '. . . *changes in buffer concentration duplicate the result of change in pairing-ion concentration, and add little to the control of retention in ion-pair chromatography systems.*' If these results prove to be generally true, paired-ion optimization will need to consider just a single member of a homologous series of ion-pair reagents and a fixed buffer strength, thereby limiting the number of variables requiring consideration.

Haky *et al.*[49] used the ORM method in the selection of the optimum solvent composition for the separation and detection of trace amounts of the *threo* isomer in samples of the bronchodilating drug procaterol. A cubic model was used to fit the retention data acquired from the seven experiments and the overlapping resolution maps were generated on a popular microcomputer.

The ORM method was used by Tsuneyoshi *et al.*[50] in designing an optimum mobile phase composition for the reversed-phase separation of eight anti-epileptic drugs. Using the least-squares method, optimum quadratic surfaces were fitted to the experimental resolution ratio data in the three-dimensional space constructed from solvent selectivity triangles and solvent strength.

A hybrid method which combined simplex lattice designs, overlapping resolution mapping and the sequential simplex procedure was employed by Weyland *et al.*[51] to optimize the reversed-phase separation of saccharin, caffeine and benzoic acid. In this example, a primary objective was to optimize the analysis time whilst maintaining the overall quality of the separation. The standard ORM method does not achieve this in reversed-phase chromatography because the proportion of water in the mobile phase is not a free variable. To overcome this limitation whilst retaining the simplicity of the planar overlapping resolution maps, the mobile phase constituents were restricted to methanol, acetonitrile and water. A fixed step-size sequential simplex procedure (see Chapter 6) was used to locate approximately the region of the factor space (in this case a triangle) which was assumed to contain the optimum conditions. A simplex lattice design was then used to determine the nature of the response surface in this constrained region, fitting a quadratic model to the experimental data. The optimum analysis time was found by visual examination of the resolution map and also by using non-linear programming.

4.4.3. Software

Not surprisingly, since much of the early ORM software formed part of a proprietary chromatograph, literature on software for the ORM method is not very extensive. Nevertheless, two papers that are of considerable interest have been published by Issaq and co-workers. The first[52] contains two listings in FORTRAN. Program one (PEAKIN) takes care of resolutions, including where peak crossovers occur, and prepares a data file of compositions and resolutions ready for use by the second program. This second program is an SAS (Statistical Analysis System)[53] routine. SAS packages are available on many of today's mainframe and super-mini computers. The necessary contour plots are produced and the optimum composition is predicted. Issaq and co-workers demonstrated good agreement with experimental observations for the program. The second paper[37] on this subject describes a method of identifying peak retention times based on measurement of peak areas. This program could be used prior to the two described above in cases where identification of solutes by reference to standards is not possible. Unfortunately, a large computer is required for these programs but mainframe computing is not, as has been suggested,[54] a pre-requisite for the use of the ORM method.

The extension of the ORM approach by D'Agostino and co-workers does have reference to a program description.[47,55] At the time of preparing this text, the book referred to[55] had still to be published but further investigation here is definitely recommended. The capabilities and use of the program have been described[56] and compared with other schemes for computer-controlled separation optimization and the program has been released as CHEOPS,[57] a commercial off-line optimization system.

Haky *et al.*[49] utilized a FORTRAN program described by Hare and Brown[58] in the development of an AppleSoft BASIC for solving the cubic model and generating the overlapping resolution map. As this program should be readily adapted to many current 8- and 16-bit laboratory computers, it is to be hoped that listings can be made more widely available.

For those requiring a simple program for cubic model fitting to mixture design experiments, one in BASIC is included as Appendix IV. In the example given in this chapter, the program has been used with a COF as the dependent response. There is no reason why the program should not be used for the individual consideration of peak pairs and then the plots overlaid manually to provide an overlapping resolution map.

4.4.4. Summary

The ORM method is one of considerable utility and potential. It has been demonstrated to be applicable to all forms and types of chromatography used in all modes. Its uses to date in fully automated separation optimization have probably been restricted to a large degree by the limited availability of software and of suitably powerful hardware. As the 16- and 32-bit computers become commonplace in the laboratory, this situation is bound to change. A second factor likely to accelerate the use of the ORM and similar methods is the availability of on-line component identification. Diode-array detection, both UV and fluorescence, coupled to suitable peak deconvolution methods, could obviate the need for concurrent chromatographing of standards. Who knows, with such technology it may be possible to optimize selectively mixtures in the presence of complex interferences.

Advantages of the ORM approach

1. As few as seven experiments required.
2. A comprehensive picture of the system can be obtained.
3. Applicable to most forms and types of chromatography.
4. Re-optimization probably not required for a new column, since the information to hand will indicate the direction to be taken.

Against the ORM approach

1. For full application, solutes must be identified.
2. Where solutes are being identified, the number of experiments required increases in proportion to the number of solutes.
3. Software is not readily available.
4. Can be very slow for complex mixtures needing solute identification.
5. High-quality graphics or plotting routines are needed to get full benefit of visual aspects.
6. Global optimum is not necessarily identified.
7. No optimum may be found for complex samples as all contours may overlap completely.
8. Some versions may excessively restrict the area of factor space which they explore as only a triangular plane is considered.

4.5. STEP-SEARCH DESIGNS

It has already been seen that optimization can be approached by using a triangular variable space—some of the best examples use the solvent selectivity triangle to define the extremes. Many schemes then require that solutes be identified, models fitted or search directions established for the full realization of the method. There is an alternative approach that can be tried which makes no prior assumptions. The method is extremely simple, does not require complex instrumentation and is trivial to automate in its basic form.

If a three-solvent delivery system is available, it is capable of providing mobile phases that are described by a triangle whose vertices represent limiting binary compositions. In the ORM approach and simplex lattice design method, these limiting conditions will have been selected to provide approximately equal elution times. Statistical designs are used to define where experiments should be conducted. Unfortunately, the models that may then be fitted are not true representations and the location of the optimum conditions may require iterative refinements. One possible way to approach the optimization without making any prior assumptions about solute behaviour, or identity, is simply to explore the whole of the variable space according to some pre-defined pattern. Such an approach is indicated in Figure 4.20. The advantages of automation will be self-evident in that a computer can calculate the best route for minimum equilibration times and can take full control of the chromatograph. The

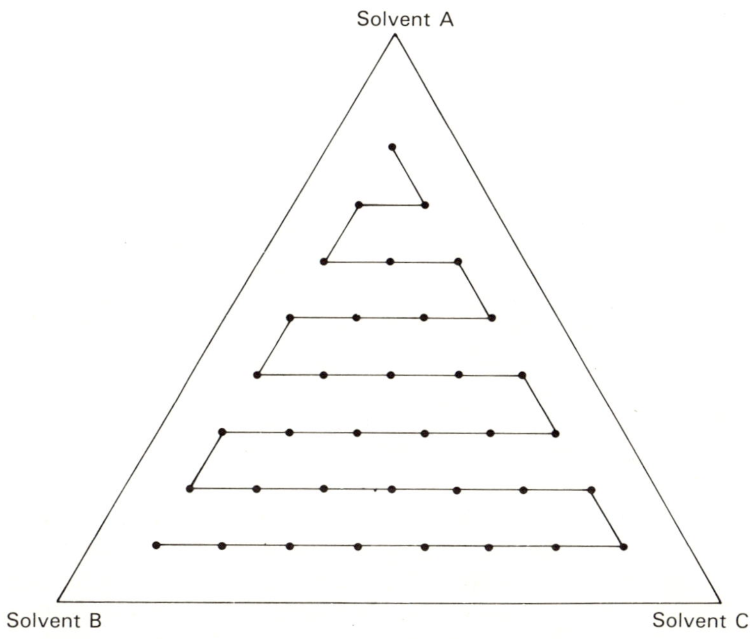

Figure 4.20. Experimental design of step-search optimization (PESOS)[59]

estimation of the optimum conditions can be undertaken in many ways. The simplest approach would be to calculate an optimization criterion for every chromatogram and then select those conditions which give rise to the best result. Perhaps several criteria can be calculated for the same separation, giving the operator even more scope for selection. Again, computer enhancement is an obvious possibility since the results could be summarized on a visual display, perhaps in colour, indicating the most promising regions for further investigation.

The software requirements for this approach are minimal and should be well within the capabilities of most amateur programmers, provided the hardware and interfaces are available. Full computer graphics are not essential, although a colour display has formed the basis for one commercially available system.[59] This approach has much to commend it as it is conceptually simple, easy to operate and does not require the success of complex algorithms. Further, there is no requirement for on-line peak identification and the choice of solvents is not restricted in number or in nature. One should beware of a major limitation, however. Supposing it is decided to investigate the whole of a triangular space as depicted in Figure 4.20 (i.e. three solvents) and let it be assumed that 10% increments are considered to be appropriate. Sixty-six experiments will be required, and the number increases exponentially as the increments get smaller! It will not usually be necessary to investigate the whole of the space, as hopefully prior knowledge can be used to restrict the search area or at least the space can be restricted as a consequence of data gained from a preliminary gradient run and the calculation of equivalent solvent strength compositions.

4.6. CONCLUSIONS

Formal statistical designs have much to commend them and this is reflected in the space devoted to them in this chapter. They represent some of the first techniques to achieve full automation and they have been included in commercially available instrumentation. The great advantages of the formal designs lie mainly in the relative ease of the data analysis, as the experimentation has been well programmed. The great drawback in many of the methods is a consequence of the need to identify solutes in order to achieve the full potential of the method. Nevertheless, all these methods are well suited for full or partial automation. A stumbling block at present, hopefully one which will continue to diminish, is that commercial success of optimization schemes results in restricted availability of software.

It will be seen in future chapters that many other systematic approaches rely both on good experimental design and on rational solvent classification, so these methods do not have the monopoly here. For those eager to establish their own approaches, the simple, unguided surface exploration represents an excellent starting point. It is necessary to get the chromatograph under control, and some feedback of the separation quality, before progressing to higher things!

4.7. REFERENCES

1. L. R. Snyder, *J. Chromatogr. Sci.*, **16** (1983) 223.
2. L. R. Snyder, *J. Chromatogr.*, **92** (1974) 223.
3. J. L. Glajch, J. J. Kirkland, K. M. Squire and J. M. Minor, *J. Chromatogr.*, **199** (1980) 57.
4. L. Rohrschneider, *Anal. Chem.*, **45** (1973) 1241.
5. M. Ryba, *Chromatographia*, **15** (1982) 227.
6. L.-C. Men, H. Na and L. B. Rogers, *J. Chromatogr.*, **314** (1984) 55.
7. H. Colin, G. Guiochon and P. Jandera, *Chromatographia*, **15** (1982) 133.
8. P. Ciccioli and R. Tappa, *J. Chromatogr.*, **269** (1983) 47.
9. L. R. Snyder, J. L. Glajch and J. J. Kirkland, *J. Chromatogr.*, **218** (1981) 299.
10. J. L. Glajch, J. J. Kirkland and L. R. Snyder, *J. Chromatogr.*, **238** (1982) 269.
11. J. L. Glajch, J. J. Kirkland and W. G. Schindel, *Anal. Chem.*, **54** (1982) 1276.
12. L. R. Snyder, 'Mobile Phase Effects in Liquid–Solid Chromatography', in Cs. Horvath (Ed.), *High-Performance Liquid Chromatography, Advances and Perspectives*, Volume 3, Academic Press, New York, 1983.
13. E. L. Weiser, A. W. Salotto, S. M. Flach and L. R. Snyder, *J. Chromatogr.*, **303** (1984) 1.
14. S. Hara and S. Ohnishi, *J. Liq. Chromatogr.*, **7** (1984) 59.
15. S. Hara and S. Ohnishi, *J. Liq. Chromatogr.*, **7** (1984) 69.
16. J.-P. Thomas, A. Brun and J.-P. Bounine, *J. Chromatogr.*, **172** (1979) 107.
17. J. L. Glajch and J. J. Kirkland, *Anal. Chem.*, **55** (1983) 319A.
18. D. L. Massart, A. Dijkstra and L. Kaufman, *Evaluation and Optimisation of Laboratory Methods and Analytical Procedures*, Elsevier, Amsterdam, 1978.
19. G. E. P. Box, W. G. Hunter and J. S. Hunter, *Statistics for Experiments. An Introduction to Design, Data Analysis and Model Building*, Wiley, New York, 1978.
20. O. L. Davis and P. L. Goldsmith (Eds), *Statistical Methods In Research and Production*, 4th ed., Longman, London, 1980.
21. J. Neter and W. Wasserman, *Applied Linear Statistical Models. Regression, Analysis of Variance and Experimental Designs*, Irwin, Homewood, IL, 1974.
22. R. C. Kong, B. Sachok and S. N. Deming, *J. Chromatogr.*, **199** (1980) 307.
23. S. N. Deming, J. G. Bower and K. D. Bower, 'Multifactor Optimisation of HPLC Conditions', in J. C. Giddings, E. Grushka, J. Cazes and P. R. Brown (Eds), *Advances in Chromatography*, Vol. 24, Marcel Dekker, New York, 1984, Ch. 2.
24. M. Otto and W. Wegscheider, *J. Chromatogr.*, **258** (1983) 11.
25. W. Lindberg, E. Johansson and K. Johansson, *J. Chromatogr.*, **211** (1981) 201.
26. M. L. Cotton and G. R. B. Down, *J. Chromatogr.*, **259** (1983) 17.
27. Y. Changhou, J. L. Fasching and P. R. Brown, 8th International Symposium Column Liquid Chromatography, New York, 1984, paper 3a29.
28. A. Tulumello and J. D. Tulumello, *Comput. Chem.*, **5** (1981) 55.
29. G. H. Golub and V. Pereyre, *SIAM (Soc. Ind. Appl. Math.) J. Numer. Anal.*, **10** (1973) 413.
30. P. R. Bevington, *Data Reduction and Error Analysis for the Physical Sciences*, McGraw-Hill, New York, 1969.
31. J. Rafel, *J. Chromatogr.*, **282** (1983) 287.
32. R. D. Snee, *Chemtech*, **9** (1979) 702.
33. R. Leher, *Ind. Res. Dev.*, **25** (1983) 116.
34. P. J. Schoenmakers, H. A. H. Billiet and L. de Galan, *J. Chromatogr.*, **205** (1981) 13.
35. L. R. Snyder, J. W. Dolan and J. R. Gant, *J. Chromatogr.*, **165** (1979) 3.
36. J. L. Glajch and L. R. Snyder, *J. Chromatogr.*, **214** (1981) 21.
37. H. J. Issaq and K. L. McNitt, *J. Liq. Chromatogr.*, **5** (1982) 1771.
38. P. E. Antle, *Chromatographia*, **15** (1982) 277.

39. M. de Smet, G. Hoogewijs, M. Puttemans and D. L. Massart, *Anal. Chem.*, **56** (1984) 2662.
40. J. J. Kirkland and J. L. Glajch, *J. Chromatogr.*, 255 (1983) 27.
41. J. L. Glajch and J. J. Kirkland, *Anal. Chem.*, **54** (1982) 2593.
42. J. L. Glajch, J. C. Gluckman, J. G. Charikofsky, J. M. Minor and J. J. Kirkland, *J. Chromatogr.*, **318** (1985) 25.
43. G. M. Landers and J. A. Olsen, *J. Chromatogr.*, **291** (1984) 51.
44. I. S. Lurie, A. C. Allen and H. J. Issaq, *J. Liq. Chromatogr.*, **7** (1984) 463.
45. I. S. Lurie and A. C. Allen, *J. Chromatogr.*, **292** (1984) 283.
46. V. W. Wachter, B. Sczepanik, G. Gottmann and G. Cordes, *Pharm. Ind.*, **45** (1983) 1000.
47. G. D'Agostino, F. Mitchell, L. Castagnetta and M. J. O'Hare, *J. Chromatogr.*, **305** (1984) 13.
48. A. P. Goldberg, E. L. Nowakowska, P. E. Antle and L. R. Snyder, *J. Chromatogr.*, **316** (1984) 241.
49. J. E. Haky, A. M. Young, E. A. Domonkos and R. L. Leeds, *J. Liq. Chromatogr.*, **7** (1984) 2127.
50. T. Tsuneyoshi, A. Kawamoto and J. Koezuka, *Clin. Chem.*, **30** (1984) 1889.
51. J. W. Weyland, H. Rolink and D. A. Doornbos, *J. Chromatogr.*, **247** (1982) 221.
52. H. J. Issaq, J. R. Klose, K. L. McNitt, J. E. Haky and G. M. Muschik, *J. Liq. Chromatogr.*, **4** (1981) 2091.
53. SAS Institute, *SAS Users' Guide*. Raleigh, North Carolina.
54. H. J. Issaq, 'Statistical and Graphical Methods of Isocratic Solvent Selection for Optimal Separation in Liquid Chromatography', in J. C. Giddings, E. Grushka, J. Cazes and P. R. Brown (Eds), *Advances in Chromatography*, Vol. 24, Marcel Dekker, New York, 1984, Ch. 3.
55. L. Capasso and G. Serio (Eds), *Lecture Notes in Biomathematics*, Springer-Verlag, Berlin, 1984.
56. G. D'Agostino, L. Castagnetta, F. Mitchell and M. J. O'Hare, *J. Chromatogr.*, **338** (1985) 1.
57. CHEOPS: Chromatography Optimisation System (Gilson).
58. L. B. Hare and P. L. Brown, *J. Qual. Technol.*, **9** (1977) 193.
59. *PESOS:* Perkin-Elmer Solvent Optimisation System.

5 Window Diagrams

5.1. INTRODUCTION

There is a new classification of chemistry that has grown in its own right since the late 1970s that is dedicated to furthering the applications of mathematical and statistical methods to chemical measurements. This discipline has been given the title 'chemometrics'.[1-3] In essence, chemometrics is the application of mathematical, statistical and computational methods to analytical systems and methods (a) to identify and optimize the critical experimental parameters and conditions and (b) to maximize the amount and usefulness of data obtainable from a given analytical experiment. It is true to say that today the subject of chemometrics has graduated from its infancy to a position of maturity. The principles and techniques embraced in the chemometrics concept are being applied to many different fields of the natural sciences[3,4] and to chromatography in particular.[5] In Chapter 4, factorial and mixture designs were described and they represent the use of chemometric methods in the design of suitable experimental procedures. Another chemometric procedure that is well suited to HPLC optimization is the so-called 'window diagram'[6] technique. Window diagrams are the product of a graphical procedure that permits the chromatographer to select the best separation conditions virtually by inspection, without the requirements of extensive data reduction methods or complex algorithms. An important additional benefit of using window diagrams is that they allow more subjective, secondary criteria also to be considered. The window diagrams permit the analyst to choose conditions which satisfy separation criteria and which meet the needs of considerations such as speed, sensitivity and cost.

5.2. THEORY

The window diagram procedure was initially developed by Laub and Purnell[6-8] for the optimization of gas–liquid chromatographic (GLC) separations. This original application provides an excellent introduction to the basic principles of window diagrams and provides a good example for the transfer of the ideas to liquid chromatography.

In GLC the separation selectivity is primarily a function of the column stationary phase. One way to manipulate this selectivity is to prepare columns with mixed-bed packings. The problem arises of how to determine the optimum composition of the blended packings.

It was found empirically[7] that a linear relationship existed between the capacity factors (k') and the composition of binary mixed-bed packings (Figure 5.1). With such a linear relationship it is possible to predict the capacity factor for any solute at any column composition, given only the capacity factors for that solute in each of the pure stationary phases. This being the case, it is a simple matter to calculate the relative retentions (α values; $\alpha = k'_2/k'_1$) between all possible peak pairs over the whole range of column compositions. The next step in the procedure is to plot these α values against the composition of the stationary phase. From this plot the 'windows' can be found which indicate the optimum experimental conditions.

These steps, for a GLC separation of four hypothetical solutes, are shown in Figures 5.1 and 5.2. Figure 5.1 shows the dependences of capacity factor on composition for the four hypothetical solutes. It is assumed that these dependences follow the linear model described earlier and so experiments to measure the capacity factors were required only for the two pure stationary phases. From Figure 5.1 it is possible to calculate the selectivity factors for

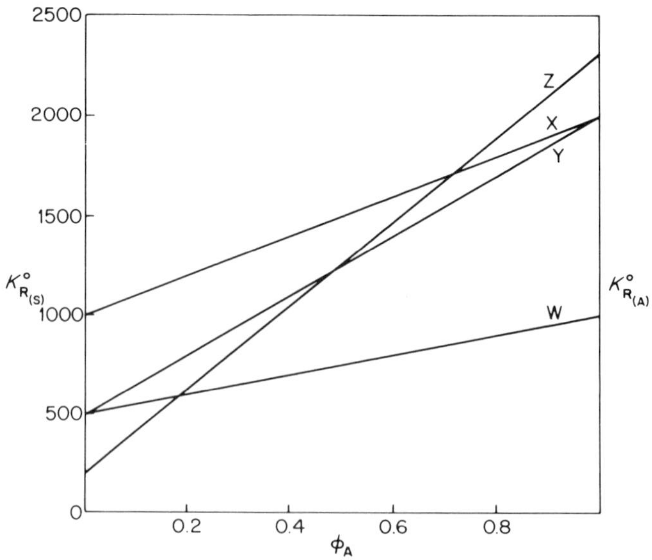

Figure 5.1. Plots of capacity factor, k_R, against stationary phase composition (ϕ_A) for four hypothetical solutes (W, X, Y and Z) separated by GLC. *Reproduced from ref. 7 by permission of Elsevier Science Publishers BV*

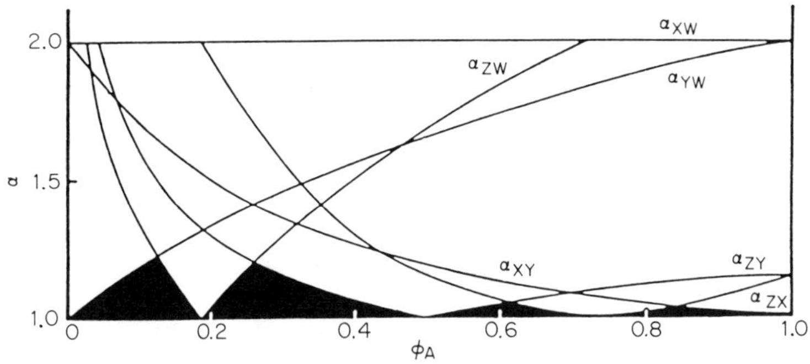

Figure 5.2. Window diagram for the solutes shown in Figure 5.1. Optimum column composition is at $\phi_A = 0.12$. *Reproduced from ref. 7 by permission of Elsevier Science Publishers BV*

all the peak pairs across the whole range of mixed packings. Figure 5.2 is the graphical representation of the selectivity, or α, data for the solutes shown in Figure 5.1. Two points should be particularly noted. First, for the four solutes under consideration, there are *six* peak pairs. In general, the number of peak pairs possible for n peaks can be found from the following equation:

$$\text{No. of peak pairs} = \frac{n!}{2(n-2)!} \tag{5.1}$$

Second, the peak separations form continuous functions, with values always greater than unity; α values greater than 1 are maintained by inverting the calculation should peak cross-overs occur; this is simply a convention but reference to Figure 5.2 quickly demonstrates the advantages of doing this. The V-shaped plots appear as a set of inverted and partially overlapping triangles, with the points of peak overlap being represented by the points of the triangles meeting the horizontal axis of the graph. The areas in which there are no overlaps can be clearly seen and these are marked as the black areas on the figure. It is these areas that are the 'windows' and within the regions that they represent separation becomes possible. The optimum separation can be achieved using the conditions that are indicated from that point on the composition axis (Figure 5.2) which corresponds to the highest point of the highest window. This point actually represents the composition ($\phi_A = 0.12$) where the best separation can be achieved between the two worst separated pairs. The top of the window also gives the value of the selectivity factor for these two worst separated pairs, since the apex of the window is formed from the intersection of two lines, each of which describes the separation between a pair of solutes.

The window diagram method is in fact a general optimization technique applicable to any situation where one or more dependent variables are to be optimized with respect to independent variables. The only proviso is that there should be a function of some form that can be defined which relates the two. Hence the procedure can be applied to HPLC just as well as to GLC.

The prerequisite for the efficient application of window diagrams to HPLC is that there should be a model which can be used to describe the relationship between the experimental variables, such as mobile phase composition, pH or column temperature, and the retention times of the individual peaks.

5.3. TWO-DIMENSIONAL MODELS

Originally developed as an essentially graphical technique, the window diagram method is most easily applied to the optimization of a single variable. It will be seen, however, that the method is certainly not restricted to just one variable. The advantages of considering a single variable are that the development of suitable models becomes simplified and the resulting window diagrams are easier to interpret, allowing the easy determination of optimum conditions by inspection.

5.3.1. Separation Temperature

To begin with, one of the very simplest of models in HPLC will be considered which describes the effect of temperature on capacity factors. Temperature has many effects on separations[9-11] and is often neglected in separation development. Mobile phase viscosity decreases and diffusion coefficients increase with increase in temperature, giving an increase in column efficiency. Capacity factors are also dependent on the separation temperature, which results in temperature being a worthwhile variable to investigate for selectivity and resolution improvement, particularly if a complex separation is being attempted.

The model for temperature effects is extremely simple, following directly from thermodynamic considerations:

$$\log k' = a + b/T \tag{5.2}$$

where a and b are experimentally derived coefficients. For most cases plots of $\log k'$ against the reciprocal of the (absolute) temperature (T) are linear, although some marked deviations have been noted,[11] for example in reversed-phase separations when mobile phases containing more than 30% of tetrahydrofuran (THF) are used. Where a linear dependence can be assumed, all that is required in practice is to determine capacity factors at two temperatures. Capacity factors at all intermediate temperatures can be calculated from equation 5.2. Plots of capacity factor against temperature will give a series of straight lines, as shown in Figure 5.3 for the separation of amino acids.[12] From these linear plots, the selectivity factors (or in this case their logarithms) are easily determined and can be plotted against the reciprocal of the temperature to give a window diagram, as shown in Figure 5.4.

It is interesting to see that, from the window diagram, two conditions occur

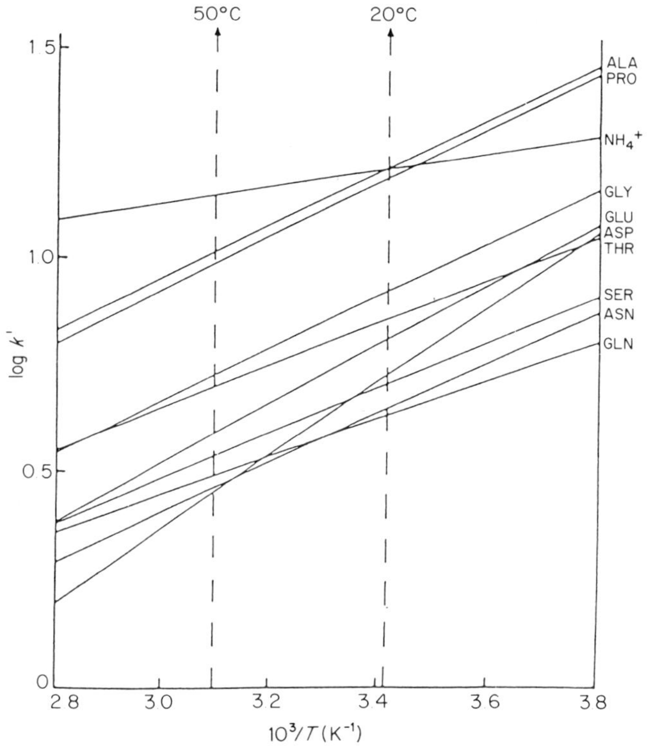

Figure 5.3. Plots of log k' against reciprocal of temperature for amino acids and ammonia with a buffered mobile phase (pH 2.6) and octyl-bonded phase. *Reproduced from ref. 12 with permission of Elsevier Science Publishers BV*

for separations of almost equal selectivity. Two windows are seen, at 55 and 15 °C, both of which give α values of about 1.06. On the face of it there is little to choose between the conditions. The fact that two optima are indicated is one of the additional advantages of the use of window diagrams. Secondary criteria, both objective and subjective, can now be introduced. For example, considerations of convenience and cost may be important, or perhaps speed or sensitivity need to be taken into account. At the higher temperature the column efficiency will be higher: sensitivity should be increased and the separation time will be reduced. If time is important, the higher temperature will be the condition to use. There may, however, be practical or economic reasons which detract from the higher temperature separation. In such cases, then the lower temperature might be selected—it does not require the use of a column heating device. Unfortunately, it may be that the ambient temperature exceeds 15 °C, which will require the use of a cooling system to achieve the separation.

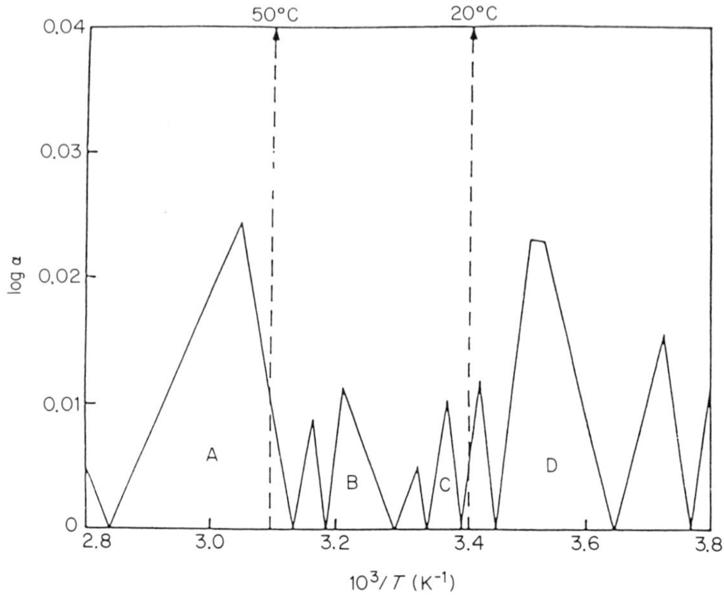

Figure 5.4. Window diagram for the solutes shown in Figure 5.2. Optimum regions are described by windows A and D. *Reproduced from ref. 12 by permission of Elsevier Science Publishers BV*

The calculation and plotting problems, particularly for non-automated temperature optimization, can be eased[13] by plotting retention time against temperature on a semi-logarithmic scale. The window diagram plots can also easily be constructed by plotting retention time ratios against temperature.

5.3.2. Mobile Phase Composition

When it comes to the consideration of window diagrams for the optimization of solvent composition, then the establishment of a valid model becomes slightly more difficult and the choice of models expands considerably. Fortunately, only a semi-empirical model is required, preferably the simpler the better. It is unfortunate that, for most mobile phases, there is not a simple, definitive linear relationship between solute capacity factors and mobile phase composition. Over a limited composition range for binary mobile phases, in ***reversed-phase*** separations, a linear relationship between log k' and mobile phase composition has been shown[14,15] for certain organic modifiers. This relationship has the form:

$$\log k' = a + b\phi_B \tag{5.3}$$

where ϕ_B is the volume fraction of the stronger solvent and *a* and *b* are the experimentally derived regression coefficients.

For reversed-phase separations, equation 5.3 gives good results over a reasonable range of compositions and is certainly worth investigating for 10–80% methanol or acetonitrile mixtures with water. Being a linear relationship, equation 5.3 has the merits of requiring only two determinations to be able to estimate the values of *a* and *b*. The linearity of the relationship is much poorer with tetrahydrofuran/water mixtures since many solutes exhibit markedly curved log k' versus composition dependences for THF and equation 5.3 cannot be unreservedly recommended for this solvent.

Many solutes also exhibit markedly curved log k' versus composition plots even with limited mixtures of methanol or acetonitrile and water. This then necessitates the development of a more complex model to describe the chromatographic behaviour. All that may be required to account for the observed curvature is the addition of a quadratic term[16] to give

$$\log k' = a\phi^2_B + b\phi_B + C \tag{5.4}$$

Even this quadratic model may turn out to be inadequate for some solutes and a more complex one will be called for. One such model which has been developed[17] to describe retention in reversed-phase chromatography with binary mobile phases (and which is reported[17] also to be applicable to normal-phase chromatography) is

$$\frac{1}{k'_{(M)}} = \phi_A \left\{ \frac{1}{k'_{(A)}} + \frac{b\phi_S}{1 + b'\phi_S} \right\} + \frac{\phi_S}{k'_{(S)}} \tag{5.5}$$

where $k'_{(M)}$ is the capacity factor at any mobile phase composition and ϕ is the volume fraction of the solvent with subscripts A and S referring to pure mobile phases, S being pure water in reversed-phase chromatography. The parameters b and b' are system constants derived from analysis of the experimental data.

In the cases investigated by its proponents, equation 5.5 has been found to account for the non-linear behaviour of many solute/solvent systems, although originally only limited ranges of solvent composition (i.e. 20–45% THF) were fully investigated. To take into account even the very marked curvatures frequently seen at the extreme ends of the mobile phase composition ranges, it was found necessary to extend equation 5.5 and add a second term:

$$\frac{1}{k'_{(M)}} = \phi_A \left\{ \frac{1}{k'_{(A)}} + \frac{b\phi_S}{1 + b'\phi_S} + \frac{c\phi_S}{1 + c'\phi_S} \right\} + \frac{\phi_S}{k'_{(S)}} \tag{5.6}$$

It is unfortunate that the values of b, b', c and c' still have to be deduced from experimentally acquired data. The full significance of these parameters remains to be elucidated. What is so far apparent is that b reflects to an extent the degree of interaction of solutes with the mobile phase additive that is sorbed on to the stationary phase surface. The value of b' does not appear to be correlated with any perceived chromatographic or solute property.

The advantages that accrue from use of the equation 5.6 are hopefully those of its generality. A greater amount of experimental data will be needed (in comparison with earlier models) at a number of mobile phase compositions such that the values of the constants can be determined. The two extremes of the mobile phase range should be included but in practical terms one can use a confined region. For example, 30–60% methanol might be considered to encompass the range under study and would then represent S and A in equation 5.6. If the model is obeyed, retentions at any other composition within the range can be obtained by simple interpolation. However, if only a limited solvent range is being investigated then equation 5.3 is recommended as a first model since the complexity of equation 5.6 detracts from its easy use.

For ***normal-phase chromatography***, a modification of equation 5.3 may be used (equation 5.7), where the logarithm of the molar concentration of the stronger solvent B is used in the calculation.[14,18,19]

$$\log k' = c - d \log X_s \tag{5.7}$$

where X_s is the molar fraction of the stronger solvent and c and d are constants ($c = \log k'$ for a pure stronger solvent). The constant c is related to the characteristics of the stationary phase and d is dependent on the solute structure.

Equation 5.7 may prove to be entirely adequate, particularly if the concentration of the polar component of the mobile phase is sufficiently high to deactivate the adsorbent surface. If this situation should pertain, then equation 5.8 will also apply[20] and may have advantages over equation 5.7.

$$1/k' = a + b\phi_B \tag{5.8}$$

where ϕ_B is the molar fraction of the modifying solvent. As experienced with reversed-phase chromatography, non-linear composition–retention relationships are sometimes observed and thus a more flexible model is required. This need can be satisfied for non-ionic normal-phase separations also[20] by use of equation 5.5. Unlike the reversed-phase situation, it is not usually necessary in normal-phase chromatography to add a second term to equation 5.5 (i.e. giving equation 5.6).

Another variable well suited to window diagram optimization is that of the ***pH*** of the mobile phase. The following model was developed[21,22] to relate pH to peak retention times:

$$t_r = \frac{t_{HA} + t_A \times 10^{(pH - pK_a)}}{1 + 10^{(pH - pK_a)}} \tag{5.9}$$

where t_{HA}, t_A and pK_a refer to the retention times of the acidic and basic forms of the compound and its dissociation constant, respectively.

The validity of this model has been demonstrated with weak aromatic acids for which pK_a values were available. If such data are not available, then experiments conducted at at least three different pH values will be required to permit the estimation of the relevant parameters of the model. The fitted parameters will, of course, be different for each solute under investigation; similarity immediately indicates potential problems in obtaining an adequate separation where pH is the only variable being manipulated. The window diagram can be constructed by calculating and plotting resolutions or selectivity factors as before.

5.4. THREE-DIMENSIONAL MODELS

Window diagram principles are not restricted to the consideration of just a single independent variable. This is fortunate, since it is the number of different variables available to the chromatographer that is both the attraction of HPLC and the reason for the complexities of developing a separation. It was not long, therefore, before the two-dimension (one independent variable) window diagram approach was extended to three dimensions (two independent variables). Unfortunately, visualizing mathematical data in three dimensions is difficult. If two or more variables are going to be considered simultaneously then computer help is an almost certain requirement, either for plotting the data or for searching through them to locate the optima. Of course, this extension to more variables also requires that suitable models be available to work across the dimensions, particularly where the variables are interdependent.

There are no major reasons why the models already described cannot be combined for use in three dimensions, for example binary mobile phase composition and temperature.[13] Algorithms are available which can be used to model the retention behaviour of solutes in reversed-phase chromatography as a function of both temperature and binary mobile phase composition:[10]

$$\log k' = A_1\phi(1 - T_c/T) + A_2/T + A_3 \tag{5.10}$$

where ϕ is the volume fraction of the organic modifier, T the absolute temperature and T_c the compensation temperature (the compensation temperature is usually constant in reversed-phase chromatography; a value of 625 K has been quoted.)[10]

Most chromatographers are likely to find that it will be necessary to consider three-component mobile phases. No simple model currently exists for ternary mobile phases but an extended form of equation 5.3 could be considered for a first approximation:

$$\log k' = a + b_1\phi_B + b_2\phi_C \tag{5.11}$$

where C refers to the second modifier. This modifier does not necessarily have to be restricted to a second organic solvent and could be a ligand or counter ion, etc.

Once again this simple approach may fail as a result of non-linear behaviour. If this is the case, then additional second-order terms need to be added to equation 5.11. A model that has been found to describe three-component mobile phases adequately includes second-order terms as well as a term to account for ternary solvent blending effects:[23]

$$\log k' = B_1\phi_1 + B_2\phi_2 + B_3\phi_3 + B_{12}\phi_1\phi_2 + B_{13}\phi_1\phi_3 + B_{23}\phi_2\phi_3 + B_{123}\phi_1\phi_2\phi_3 \quad (5.12)$$

Equation 5.12 represents a fairly complex description of solute/solvent behaviour. Careful experimental planning is required to enable the polynomial regression parameters to be obtained. It is recommended that a formal experimental design be used, such as the simplex lattice design described in Chapter 4. Having derived the capacity factor relationships, it is possible to prepare the window diagram. This is easier to say than do, and advanced regression programs and computer plotting routines are virtually indispensable.

Should even equation 5.12 fail to suffice, then for methanol, acetonitrile and tetrahydrofuran mixtures with water, equation 5.13 can be tried:[16]

$$\ln k' = C + A_M\phi^2{}_M + B_M\phi_M + E_M\phi_M{}^{0.5} + A_A\phi^2{}_A + B_A\phi_A + E_A\phi_A{}^{0.5} + A_T\phi^2{}_T + B_T\phi_T + E_T\phi_T{}^{0.5} \quad (5.13)$$

This model has been shown to account for dramatic curvatures in log k' versus composition plots, particularly at high water compositions, but again much experimentation will be needed to obtain good estimates of the regression coefficients.

When examining the effects of two (or more) variables on the selectivity of the separation, the term 'window diagram' is perhaps slightly misleading. What is more correctly being done is that a three-dimensional space is being examined for a solvent composition which gives the maximum separation between the worst separated solute pair. An alternative name for diagrams depicting this is 'minimum alpha plots' (MAPs).[23] This terminology assumes that it is the α values that are being considered, but the window diagram, or MAP, technique does not restrict itself to the use of α values only. It is worth considering the relationships that can be used with the window diagram method, as there are many.

5.5. WINDOW DIAGRAM PLOTTING PARAMETERS

When used originally for gas chromatography, window diagrams were plotted considering relative volatilities (α). The direct analogy for HPLC is the selectivity factor. The selectivity factor (α value) is simply the ratio of the two capacity factors of the peaks under consideration. The precise calculation of capacity factor requires a knowledge of the column void volume, which can be difficult to obtain and which is also subject to variation depending on the solvent composition.[24,25] Bearing this in mind, other values should be considered which can be used to construct window diagrams and which may ease the calculation problems.

Resolution between peaks is a second parameter which can be, and has been, used for constructing window diagrams but, like α values, it too has severe drawbacks. The calculation of resolution requires peak base widths, or peak half-height widths, to be calculated or estimated. Accurate values for both are not easy to determine and so, as with separation factors, problems will be experienced.

To overcome these problems, a very attractive alternative equation for window diagram construction has been developed by Jones and Wellington.[26] Applying plate theory to the separation of two solutes, they were able to define a new parameter, a separation factor, S:

$$S=(V_B-V_A)/(V_A+V_B) \tag{5.14}$$

where V_A and V_B are the elution volumes of solutes A and B, respectively. By using S values, the requirement to determine the column void volume or time is immediately removed; additionally, no peak parameters (such as base width) are required. It is also not necessary to work in units of volume since retention times can be used as direct substitutes.[13]

Another window diagram ordinate plot proposed, which does not require void time determinations, is that of retention time ratios.[13,27,28] These are probably the easiest values to calculate but with automated procedures the time and effort saved are not significant. Rather, one should perhaps use equation 5.14 since from it may be gained additional information such as the number of plates (N) required for a defined criterion of separation. The plate requirement can be found[26] from the relationship

$$2R/N^{0.5}=S \tag{5.15}$$

An additional reason for using equation 5.14 is that it produces window diagrams that parallel those produced from α or resolution plots.[17] This is particularly so when capacity factors approaching 10 are obtained, but at low capacity factors some deviations can be seen.

The direct relationship of S with α is easily seen from equation 5.17. The function S can also be defined by

$$S=\frac{k'_A(\alpha-1)}{k'_A(\alpha+1)+2} \tag{5.16}$$

Since α will always be defined as not less than unity, when $k'_A \gg 1$, equation 5.16 can be reduced to

$$S=(\alpha-1)/(\alpha+1) \tag{5.17}$$

Equation 5.17 then gives the relationship between S and α values. Thus equivalent figures are easily produced without the complications that arise from the need to deduce difficult or ill-defined values. Where low values of capacity factors are obtained, equation 5.14 has even more to commend it, because any small errors in the estimation of the column void volume will have large effects on capacity factors. These errors will result in S plots seeming to be no longer equivalent to α plots, the assumptions made in going from equation 5.16 to equation 5.17 no longer being valid.

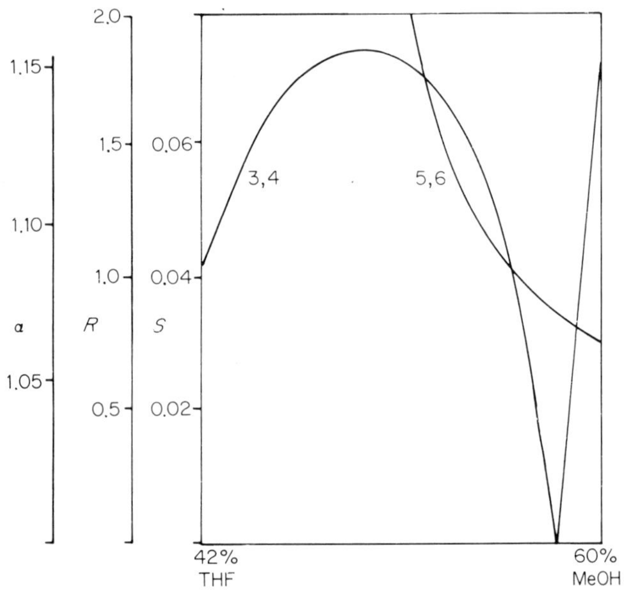

Figure 5.5. Window diagram for the six-solute test mixture showing direct equivalence of α, R and S parameters in window diagram plotting

If the separation of the six-component mixture is examined, many of these points can be illustrated. Figure 5.5 is the window diagram for the separation of the mixture when the ternary mobile phase water/methanol/tetrahydrofuran is considered. The mobile phases are produced by combining various proportions of two binary mixes, i.e. 60% methanol in water and 42% THF in water; these two mixtures provide approximately equal solvent strengths but great contrasts in selectivity. The ordinate shows three plotting parameters, i.e. resolution, separation factor and selectivity factor. The diagram is the same for all three parameters, confirming that any one may be chosen. No model has been used to fit the experimental data; instead, six experiments were conducted over the range of mobile phase compositions and the data plotted directly to give the figure shown. This is a slightly unconventional use of the window diagram method, not using a model, but it does serve to illustrate both the simplicity and versatility of the procedure.

For the test mixture, only two solute pairs actually need to be considered, 3,4 and 5,6. From Figure 5.5 it is seen that peaks 3 and 4 control the separation goal for most of the composition range except for the region from 60 to 80% of the methanol/water solvent. The meeting of the 3,4 line with the abscissa results from the peak cross-over of these two components. To achieve a target resolution of 1.5 for the worst separated pair, a wide range of compositions is indicated, from 10 to 70% of the methanol/water solvent. Note that all data were obtained with mobile phase mixtures giving approximately equal retention times (iso-elutropic), so from this plot the secondary criterion of time cannot be introduced.

5.6. APPLICATIONS

One of the first published applications of window diagrams to aid the development of HPLC separations was the optimization of mobile phase pH in the separation of benzoic acids,[21] for which the model has already been described (equation 5.9). This separation has been further investigated[22] and represents the first example of the automated optimization of a separation using

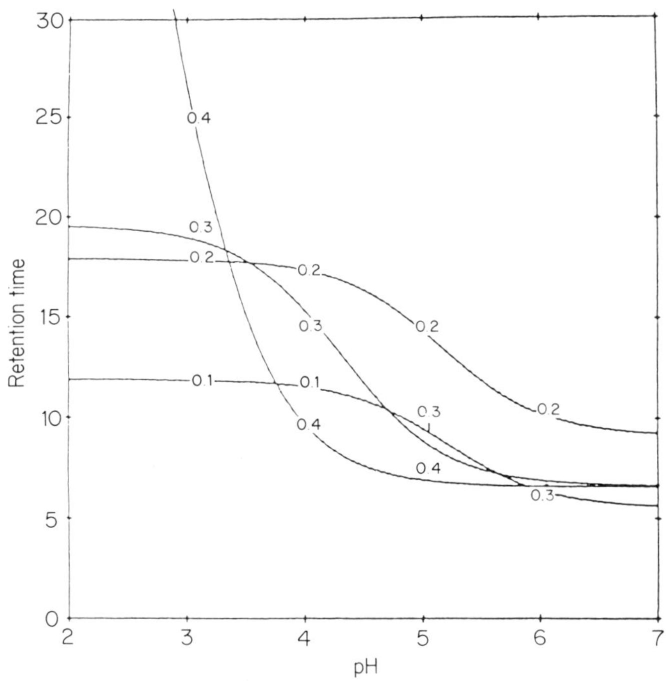

Figure 5.6. Experimental and predicted retention data as functions of pH for four carboxylic acids. Solutes: 1 = *o*-aminobenzoic acid; 2 = *p*-aminobenzoic acid; 3 = *p*-hydroxybenzoic acid; 4 = *o*-hydroxybenzoic acid.[22] *Reprinted from* American Laboratory **16***(4), 69 (1984). Copyright 1984 by International Scientific Communications, Inc*

window diagrams. Figures 5.6–5.9 show the stages in the optimization of that separation and provide a useful sequence for a fuller understanding of both the principles and application of the technique.

Extending the window diagram idea to multi-factor optimization, Sachok *et al.*[29] examined pH and the concentration of the ion-pair reagent in the optimization of the reversed-phase separation of a nine-component mixture consisting of both acids and bases. The utility of a two-factor procedure using minimum alpha plots was further investigated for optimizing the ion-pair reagent concentration and methanol content in the reversed-phase separation of substituted anilines.[30] Nine factor combinations were required to provide sufficient experimental data to permit the fitting of a mechanistic model.

Also considering mobile phase pH as one variable, but now including ionic strength and organic modifier (methanol) content, Otto and Wegscheider[31] examined a seven-component mixture of organic acids, amino acids and dipeptides. Part of the work was concerned with the development of a new model and multi-dimensional window diagrams were plotted. The global optimum was actually located by a computerized grid search of about 7000 experimental conditions! What is of particular interest in this application is the number of local optima that are revealed when the search increments are made relatively

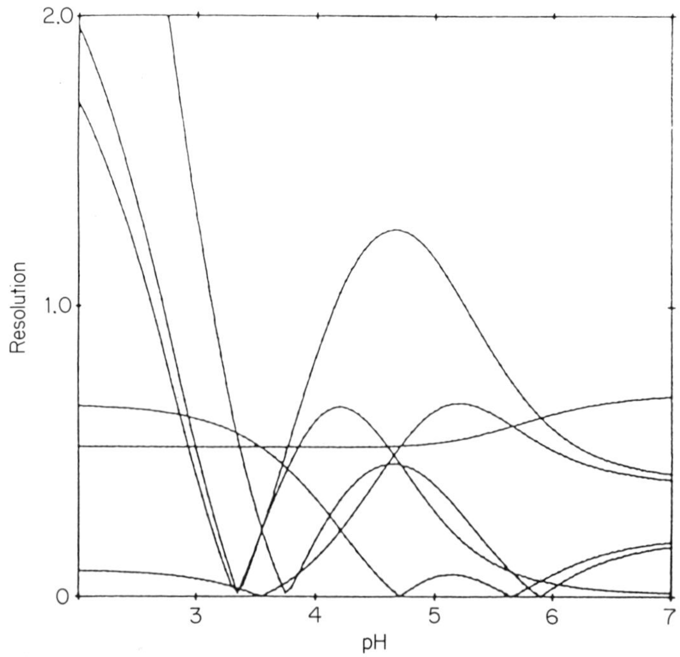

Figure 5.7. Minimum resolution as a function of pH for all possible peak pairs of the four solutes of Figure 5.6.[22] *Reprinted from* American Laboratory, **16***(4), 69 (1984). Copyright 1984 by International Scientific Communications, Inc*

small, indicating the problems that face even the most experienced chromatographer when trying to optimize a multi-factor separation.

Another investigation into the separation of aromatic acids was carried out by Patel *et al.*[32] Eluent composition, pH and temperature were investigated in turn and window diagrams constructed using the separation parameter *S*. By considering variables sequentially, it was possible to use simple window plots to indicate the optimum conditions resulting from the variation of many parameters.

The reversed-phase separation of PTH-amino acids has been investigated by Noyes[13] and this application provides a clear example of how simple and effective window diagrams can be, even for complex separations and when considering many variables. Solvent strength (acetonitrile concentration) was investigated at constant temperature and temperature was investigated at constant solvent strength. Window diagrams were simply constructed by plotting the ratios of retention times against temperature or against percentage of organic modifier (Figures 5.10 and 5.11). From Figure 5.11 it can be seen that the optimum conditions are to be found at 37.8% acetonitrile and the separation that composition produced is shown in Figure 5.12. Further use of the data enabled a gradient separation to be devised.

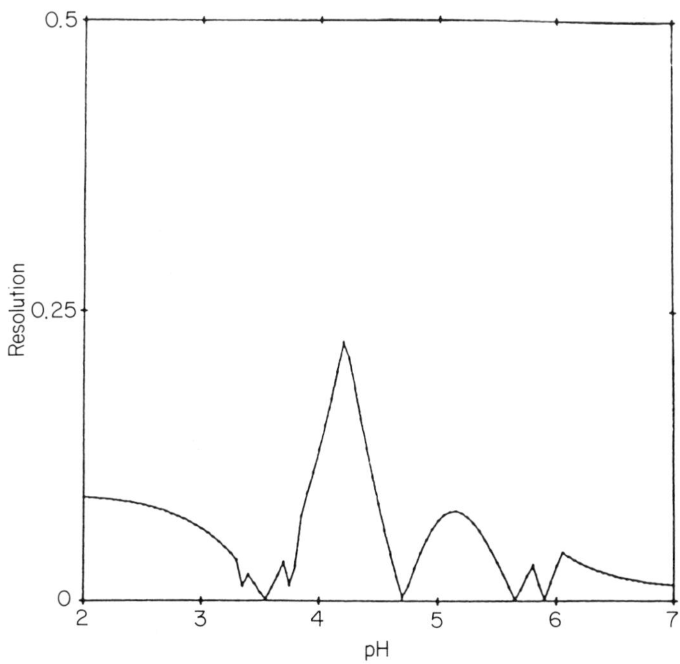

Figure 5.8. Window diagram based on the results shown in Figure 5.7[22] *Reprinted from* American Laboratory, **16***(4), 69 (1984). Copyright 1984 by International Scientific Communications, Inc*

As part of an investigation into the retention model described by equation 5.5, Hsu *et al.*[17] investigated the separation of nine phenolic solutes. The relationship between different window diagram ordinate parameters, namely α values, resolution and separation factors (S), were also discussed. It has already been seen in Figure 5.5 that there is a direct correlation between these parameters, demonstrating the value of the factor S for window diagram plotting.

There is growing interest in the use of indirect photometric methods for the detection of solutes which do not possess useful chromophores. In particular, this technique is being exploited for the analysis of organic and inorganic anions. Jenke[33] developed a model to describe the behaviour of anions as a function of mobile phase pH and ionic strength for ion-exchange separations. Window diagrams were used to illustrate this behaviour and to permit optimum conditions to be selected.

Three-dimensional window diagrams, or minimum alpha plots, were used by Weyland *et al.*[23] for the optimization of the separations of three mixtures of sulphonamides with ternary mobile phases. The plots are complex, particularly where peak cross-overs occur, but they give large amounts of information. One such plot is shown in Figure 5.13 and it can be seen that conditions for either optimum resolution or optimum speed, given a resolution constraint of 1.25, are surprisingly easy to find. Good computer graphics software is required, however!

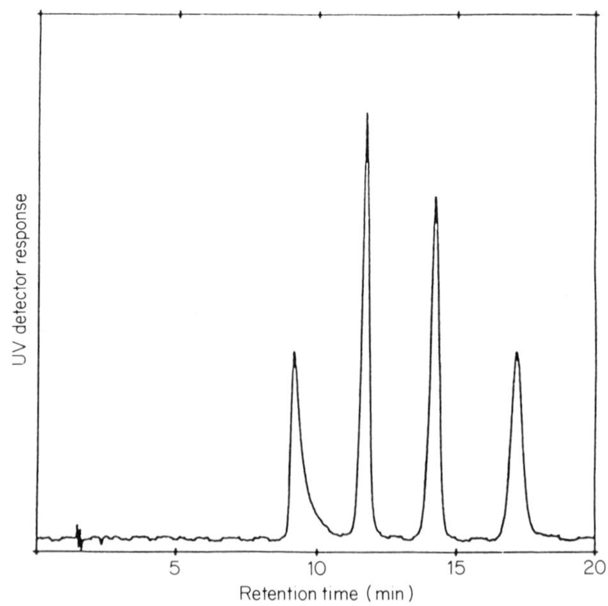

Figure 5.9. Optimum chromatogram as predicted by the window diagram of Figure 5.8, i.e. pH = 4.18.[22] *Reprinted from* American Laboratory, **16***(4), 69 (1984). Copyright 1984 by International Scientific Communications, Inc*

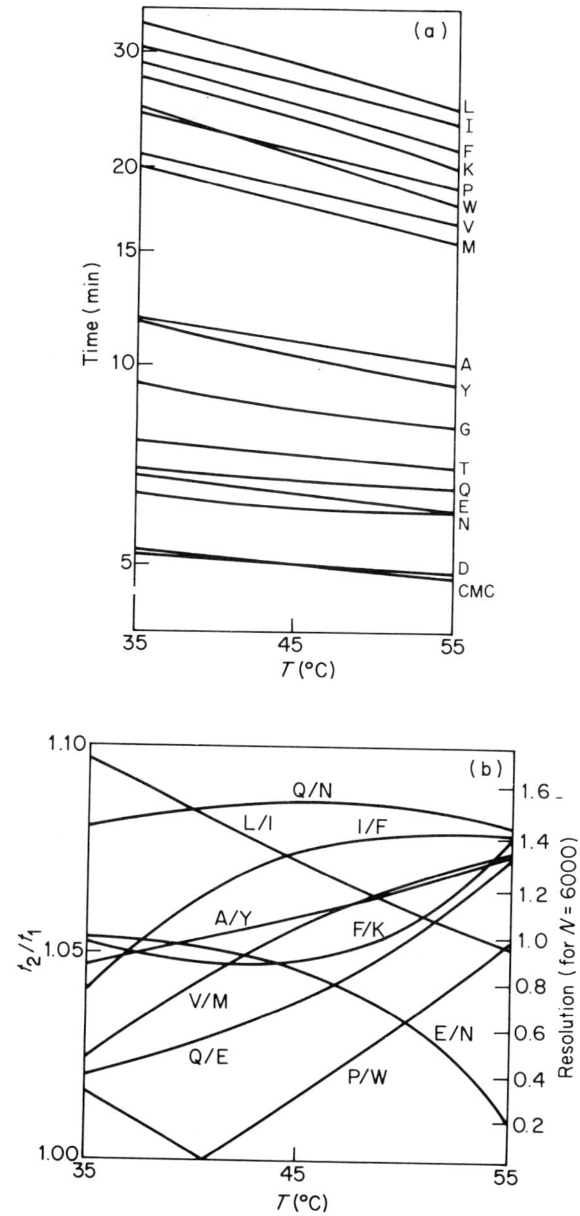

Figure 5.10. (a) Plot of logarithm of retention time against temperature for PTH-amino acids on a Partisil ODS-2 column (50 cm × 4.6 mm i.d.). Mobile phase, 0.01 M sodium acetate in acetonitrile/water (40:60); measured pH, 4.9. Solutes: D = Asp; CMC = *S*-carboxymethyl-Cys; N = Asn; E = Glu; T = Thr; G = Gly; Y = Tyr; A = Ala; M = Met; V = Val; P = Pro; W = Trp; K = Lys; F = Phe; I = Ile; L = Leu. (b) Window diagram of t_2/t_1 against temperature for the data shown in (a). *Reproduced from ref. 13 by permission of Elsevier Science Publishers BV*

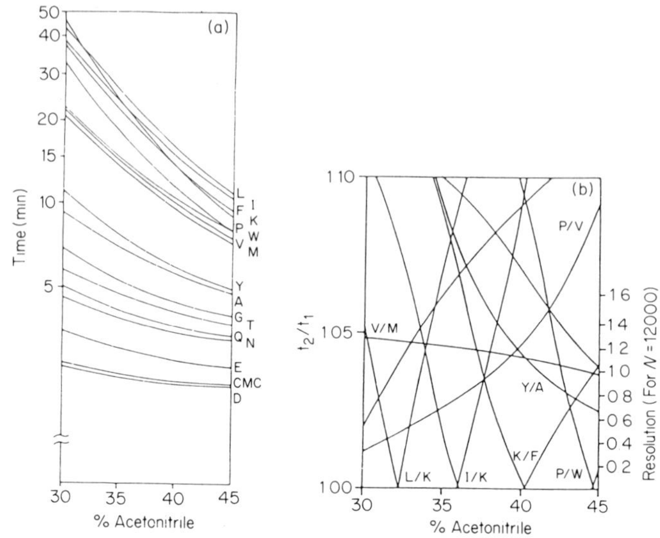

Figure 5.11. (a) Plot of logarithm of retention time against percentage of acetonitrile on an Ultrasphere-ODS column (25 cm × 4.6 mm i.d.). Mobile phase, 0.01 M sodium acetate (pH 4.9)/acetonitrile; temperature, 55°C. (b) Window diagram of t_2/t_1 against percentage of acetonitrile for the data shown in (a). *Reproduced from ref. 13 by permission of Elsevier Science Publishers BV*

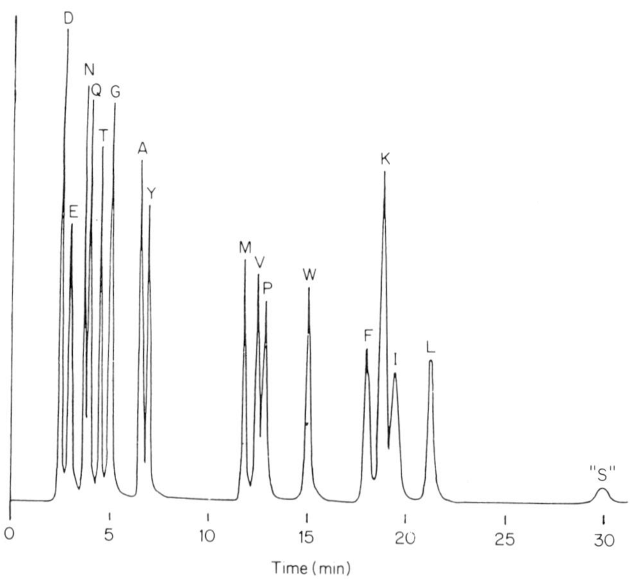

Figure 5.12. Isocratic separation of PTH-amino acids on an Ultrasphere-ODS column (25 cm × 4.6 mm i.d.). Mobile phase, 0.01 M sodium acetate (pH 4.9)/acetonitrile (62.2:37.8); temperature, 55°C. "S" = degradation product of PTH-Ser; other solutes as in Figure 5.10. *Reproduced from ref. 13 by permission of Elsevier Science Publishers BV*

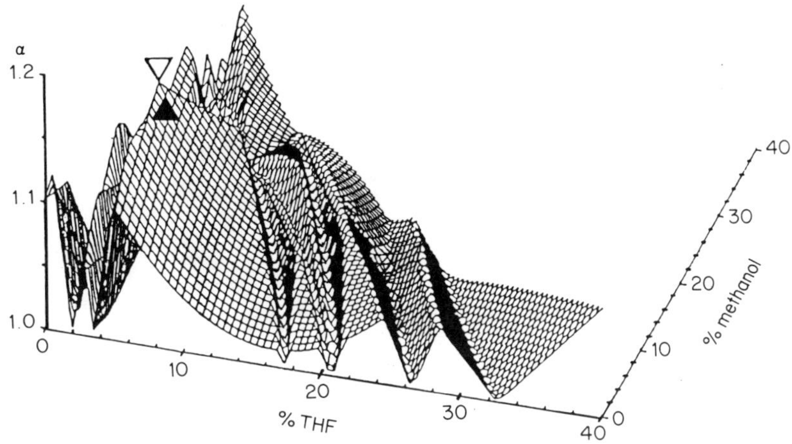

Figure 5.13. Minimum alpha plots (▽ , optimum composition; ▼ , composition that yields the fastest separation while maintaining $R > 1.25$ for a column of 2600 theoretical plates) for the separation of five sulphonamides using reversed-phase chromatography with water/methanol/THF mixtures. The complicated surface arises as a consequence of multiple peak crossings. *Reproduced from the* Journal of Chromatographic Science, *ref. 23, by permission of Preston Publications, Inc*

The applications listed above are by no means exhaustive but are intended to give an idea of the diversity of separation types to which the idea has been applied. Also, the applications chosen give further insights into just how methods can be applied to practical problems. A more comprehensive list of applications is provided in Table 5.1, with literature references.

5.7. SOFTWARE

It is unfortunate that so little software is available in the literature to aid in the use of window diagrams for either manual or automated optimization. A reference to totally automated optimization on a commercially available chromatograph[22] does not give sufficient information to enable the required program to be developed. One listing, in AppleSoft BASIC, is available[42] to consider window diagrams utilizing the retention model described by equation 5.5. While not describing a system for total automation, there is little doubt that the program could form the kernel of such a system. A further reference to software, particularly for the plotting of three-dimensional diagrams, has appeared[43] but again no details are provided.

TABLE 5.1. Window Diagram and Minimum Alpha Plot (MAP) Applications in HPLC

Application*	Variables	Ref.
R-P separation of benzoic acids	Mobile phase pH	21, 22
R-P separation of cinnamic acids	Mobile phase pH	34
Normal-phase separations of barbiturates, aromatic bases and phenothiazines	Mobile phase polarity	35
R-P separation of acids and bases	pH and ion-pair concentration	29
R-P separation of aromatic acids	Mobile phase pH	26
R-P ion-pair separation of 2,6-disubstituted anilines	Ion-pair and methanol concentrations	30
R-P ion-pair separation of phencyclidine mixtures	Methanol content	36
R-P separation of dibasic acids, bases and peptides	pH and methanol content	31, 37, 38
R-P separation of organic acids	Mobile phase pH	39
R-P separation of aromatics	Methanol content	27, 28
R-P separation of PTH-amino acids	Temperature, acetonitrile content	13
R-P separation of nine phenols	THF content	17
R-P separation of sulphonamides	Ternary mobile phase compositions	23
R-P separation of aromatic acids	Temperature, pH, methanol content	32
Ion chromatography of organic and inorganic anions	pH, buffer strength	33, 40
R-P separation of components in cough/cold formulation	Mixed ion-pair modifiers	41

*R-P = reversed-phase.

5.8. SUMMARY

Window diagram techniques provide a useful method for HPLC optimization but they ideally require that some mathematical description of the variation of the dependent parameters is available. Faced with a separation to optimize, it is recommended that the simplest model (and hence the least number of experiments) be investigated first. Should the simple approach fail, then a more refined, and inevitably complex, model can be introduced.

Advantages of Window Diagrams

1. The *global optimum* should be located, depending on criteria.
2. Multiple local optima should be visible, allowing the user to select conditions that, whilst not providing the best possible separation, permit other factors to be taken into consideration.
3. The application, particularly to two-factor optimization, is simple.
4. A good understanding of the response surface can be acquired.

5. Following model fitting, an optimum separation should be achieved with the first subsequent chromatogram.
6. The technique is applicable to all forms of chromatography.

Against Window Diagrams

1. Solutes must be identified.
2. Retention behaviour should be fitted to a suitable model (which does not necessarily have to be complex).
3. For complex models, large amounts of experimental data may be needed to allow coefficients to be calculated.
4. Where more than one variable is being considered simultaneously, the visualization of the window plots becomes difficult. For three interdependent variables, computer searching will be required.
5. The complexity of computer software rises in proportion to the complexity of the model and the number of variables being considered.

5.9. REFERENCES

1. B. R. Kowalski, *Trends Anal. Chem.*, **1** (1981) 71.
2. D. L. Massart and G. Hoogewijs, *Pure Appl. Chem.*, **55** (1983) 1861.
3. D. Betteridge, *Lab. Pract.*, October (1983) 13.
4. 'Fundamental Reviews in Analytical Chemistry:' a chemometrics review is published bi-annually, e.g. M. F. Delaney, *Anal. Chem.*, **56** (1984) 261R.
5. S. T. Balke, *Quantitative Column Liquid Chromatography—a Survey of Chemometric Methods*, Journal of Chromatography Library, Vol. 29, Elsevier, Amsterdam, 1984.
6. R. J. Laub, *Int. Lab.*, May/June (1981) 16.
7. R. J. Laub and J. H. Purnell, *J. Chromatogr.*, **112** (1975) 71.
8. R. J. Laub in T. Kuwana (Ed.) *Physical Methods in Modern Chemical Analysis*, Vol. 3, Academic Press, New York, 1983, Ch. 5.
9. L. R. Snyder, *J. Chromatogr.*, **179** (1979) 167.
10. W. R. Melander, B.-K. Chen and Cs. Horvath, *J. Chromatogr.*, **318** (1985) 1.
11. R. J. Laub and S. J. Madden, *J. Liq. Chromatogr.*, **8** (1985) 187.
12. R. J. Laub and J. H. Purnell, *J. Chromatogr.*, **161** (1978) 49.
13. C. M. Noyes, *J. Chromatogr.*, **266** (1983) 451.
14. L. R. Snyder, *Anal. Chem.*, **46** (1974) 1384.
15. P. J. Schoenmakers, H. A. H. Billiet and L. de Galan, *J. Chromatogr.*, **185** (1979) 179.
16. P. J. Schoenmakers, H. A. H. Billiet and L. de Galan, *J. Chromatogr.*, **282** (1983) 107.
17. A.-J. Hsu, R. J. Laub and S. J. Madden, *J. Liq. Chromatogr.*, **7** (1984) 615.
18. S. Hara, M. Hirasawa, S. Miyamoto and A. Ohsawa, *J. Chromatogr.*, **169** (1979) 117.
19. S. Hara, K. Kunihiro, H. Yagamuchi and E. Soczewinski, *J. Chromatogr.*, **239** (1982) 687.
20. A.-J. Hsu, R. J. Laub and S. J. Madden, *J. Liq. Chromatogr.*, **7** (1984) 599.
21. S. N. Deming and M. L. H. Turoff, *Anal. Chem.*, **50** (1978) 546.
22. J. H. Nickel and S. N. Deming, *Am. Lab. (Fairfield)*, **16** (1984) 69.

23. J. W. Weyland, C. H. P. Bruins and D. A. Doornbos, *J. Chromatogr. Sci.*, **22** (1984) 31.
24. H. Colin, A. Krstulovic, G. Guiochon and J. P. Bounine, *Chromatographia*, **17** (1983) 209.
25. R. J. Laub and S. J. Madden, *J. Liq. Chromatogr.* **8** (1985) 173.
26. P. J. Jones and C. A. Wellington, *J. Chromatogr.*, **213** (1981) 357.
27. H. J. Issaq, G. M. Muschik and G. M. Janini, *J. Liq. Chromatogr.*, **6** (1983) 259.
28. H. J. Issaq, 'Statistical and Graphical Methods of Isocratic Solvent Selection for Optimal Separation in Liquid Chromatography', in J. C. Giddings, E. Grushka, J. Cazes and P. R. Brown (Eds), *Advances in Chromatography*, Vol. 24, Marcel Dekker, New York, 1984, Ch. 3.
29. B. Sachok, R. C. Kong and S. N. Deming, *J. Chromatogr.*, **199** (1980) 317.
30. B. Sachok, J. Stranahan and S. N. Deming, *Anal. Chem.*, **53** (1981) 70.
31. M. Otto and W. Wegscheider, *J. Liq. Chromatogr.*, **6** (1983) 685.
32. B. Patel, J. H. Purnell and C. A. Wellington, *J. High Resolut. Chromatogr. Chromatogr. Commun.*, **7** (1984) 375.
33. D. R. Jenke, *Anal. Chem.*, **56** (1984) 2674.
34. W. P. Price, R. Edens, D. L. Hendrix and S. N. Deming, *Anal. Biochem.*, **93** (1979) 233.
35. J.-P. Thomas, A. Brun and J.-P. Bounine, *J. Chromatogr.*, **172** (1979) 107.
36. L. A. Jones, R. W. Beaver and T. L. Schmoeger, *Anal. Chem.*, **54** (1982) 182.
37. M. Otto and W. Wegscheider, *J. Chromatogr.*, **258** (1983) 11.
38. M. Otto, *Z. Chem.*, **23** (1983) 204.
39. C. Calcagno and M. Forina, *Farmaco Ed. Prat.*, **38** (1983) 49.
40. D. R. Jenke and G. K. Pagenkopf, *Anal. Chem.*, **56** (1984) 85.
41. S. J. Costanzo, *J. Chromatogr.*, **314** (1984) 402.
42. R. J.Laub, *J. Liq. Chromatogr.*, **7** (1984) 647.
43. C. A. Wellington and K. Ahmad, *Anal. Proc.*, **21** (1984) 144.
44. F. H. Walters and S. N. Deming, *Anal. Chim. Acta*, 165 (1985) 361.

6 Directed Search Techniques

One key feature distinguishes the techniques to be discussed in this chapter from those discussed previously, namely the use of information gathered during the optimization procedure to direct and guide the progress of further refinement of the separation. At each step of the process, chromatograms are evaluated against pre-defined criteria and the direction of the optimization is modified, or adapted, to account for the new information accumulated from the previous separations. The values of such an approach are self-evident. At last ideal, or even predictable, behaviour of the solutes is not assumed. Adaptive methods have the ability to cope with the unexpected *during* the optimization process. Their ultimate success in locating experimental conditions which provide a truly optimized separation does, however, depend very heavily on two factors, the power of the directing algorithm and the performance of the chromatographic quality criterion that is being used. Both of these aspects will be considered in this review of the capabilities of adaptive methods for HPLC optimization.

There are three directed search methods currently available which will be discussed in this chapter. The first is a fairly empirical approach which has been specifically designed for HPLC separation optimization. It uses 'adaptive intelligence' to direct the search for optimum conditions and has been given the name OPTIM 1 by its developers.[1] The second is based on a very powerful chemometric method called the sequential simplex procedure, which is a general optimization procedure that is by no means limited to liquid chromatography. The third is a hybrid method: it combines elements of mixture designs (see Chapter 4) with a search procedure that is not unrelated to the sequential simplex procedure.

6.1. ADAPTIVE INTELLIGENCE

The Spectra-Physics company use the phrase 'adaptive intelligence' to describe the approach they designed to carry out the automated optimization of isocratic and gradient separations.[1,2] The key feature of this approach is that it is interactive—information from each separation is fed back into the process to modify the direction of optimization. This is readily achieved using hardware and software designed by Spectra-Physics as all communications problems and interfacing requirements have been taken care of by the manufacturer. The optimization system is designed to tackle both normal- and reversed-phase isocratic separations and also reversed-phase gradient separations.

6.1.1. Theory

The guiding criterion for the optimization process is a chromatographic response function which has been termed the 'EVALUATION PARAMETER XA'. This parameter is a function of the number of peaks found, the resolution of each peak pair per unit time and the total analysis time. Each of these parameters is weighted in evaluating the quality of a chromatogram with the number of peaks being most heavily weighted. The overall objective of the function is to provide the chromatogram with the maximum value for XA: this should correspond to a chromatogram giving the maximum number of peaks in the shortest analysis time. It is not necessary to identify and track each peak during the separation, although such a capability was investigated in early versions of the software development, using a combination of peak retention times and areas.

The adaptive intelligence method was developed primarily for the optimization of binary and ternary reversed-phase separations. Since calculations are to be made during the optimization procedure to estimate iso-elutropic (i.e. equal strength) solvent mixtures, the solvents to be used must be pre-specified by the chromatographer. One essential requirement is that solvent A should be the solvent strength adjuster (e.g. water). Additionally, all peaks should elute with a solvent mixture containing 90% of solvent B. A third essential requirement is that all peaks must elute in each chromatogram. This is important since the XA parameter most heavily weights the number of peaks found in each chromatogram. Any peaks which are not detected as a consequence of their not being eluted during a run, or which may be eluted in the subsequent separation, will cause severe mis-direction of the optimization.

The optimization scheme is divided into a number of modes, the basic steps being as follows:

Mode 1

1. The procedure starts with a solvent composition of 90% B.
2. The subsequent separations are carried out by decreasing the % B by 20% steps until the EVALUATION PARAMETER XA passes through a maximum value. The maximum value of XA is determined by comparing the current value of XA with the one obtained for the previous chromatogram. If the procedure passes through 50% B then subsequent steps are all halved, i.e. 10% steps are used. After each run in Mode 1 a short gradient is run to ensure that all peaks are eluted from the column.

Mode 2

3. Having passed the maximum value of XA, the solvent strength is obviously too low and the next separation is carried out with a 10% or 5% increase in the proportion of solvent B. These increments are continued until the maximum value for XA is again passed.

Mode 3

4. The increment is now reduced to 1–2% and the solvent composition is adjusted towards the optimum. When the value of XA is maximized this time, this represents the optimum binary isocratic composition.

Mode 4

5. If a ternary mobile phase is being sought, the composition of the binary solvent mixture A–C is calculated, which should give a maximum value for XA. A chromatogram is run under these conditions. This composition is modified in a similar manner to the A–B optimization until XA is maximized.
6. A chromatogram is run using a ternary mixture consisting of 50:50 by volume of the optimum binary solvents A–B and A–C.

Mode 7

7. From the two optimum binary isocratic separations and the ternary separation, the optimum isocratic separation is calculated, run and reported.

The procedure for isocratic separation optimization is illustrated in Figure 6.1. Whilst not being entirely restricted to reversed-phase chromatography, normal-phase optimization is limited to just two solvents because of the problems associated with calculating the required proportion of the third solvent.

A very similar philosophy is contained in the procedure to optimize gradient separations. This is also illustrated in Figure 6.1. The major difference between a gradient and an isocratic optimization process is that the gradient slope is changed rather than the isocratic composition to find the best A–B mixture: it is also necessary to select the initial A–B composition for the start of the gradient. The rest of the system operation is essentially the same as in the isocratic modes. Note that in gradient optimization it is only the slope of the (linear) gradient that is changed and not the shape of the gradient. However, it is probable that a linear gradient will suffice for most gradient separations and to approach the optimization in this manner is entirely in accordance with the recommendations that were outlined in Chapter 3.[3]

In the gradient optimization procedure (Mode 5), the basic steps are as follows:

1. Carry out a separation with a 70:30 isocratic A–B mixture. Check to see whether there are 2–4 peaks in the k' range up to $k'=4$. If no peaks are detected then the optimization is halted. The criterion determining the initial composition is that just 2–4 peaks should have capacity factors of less than 4. If more than four peaks are found then the solvent strength is decreased

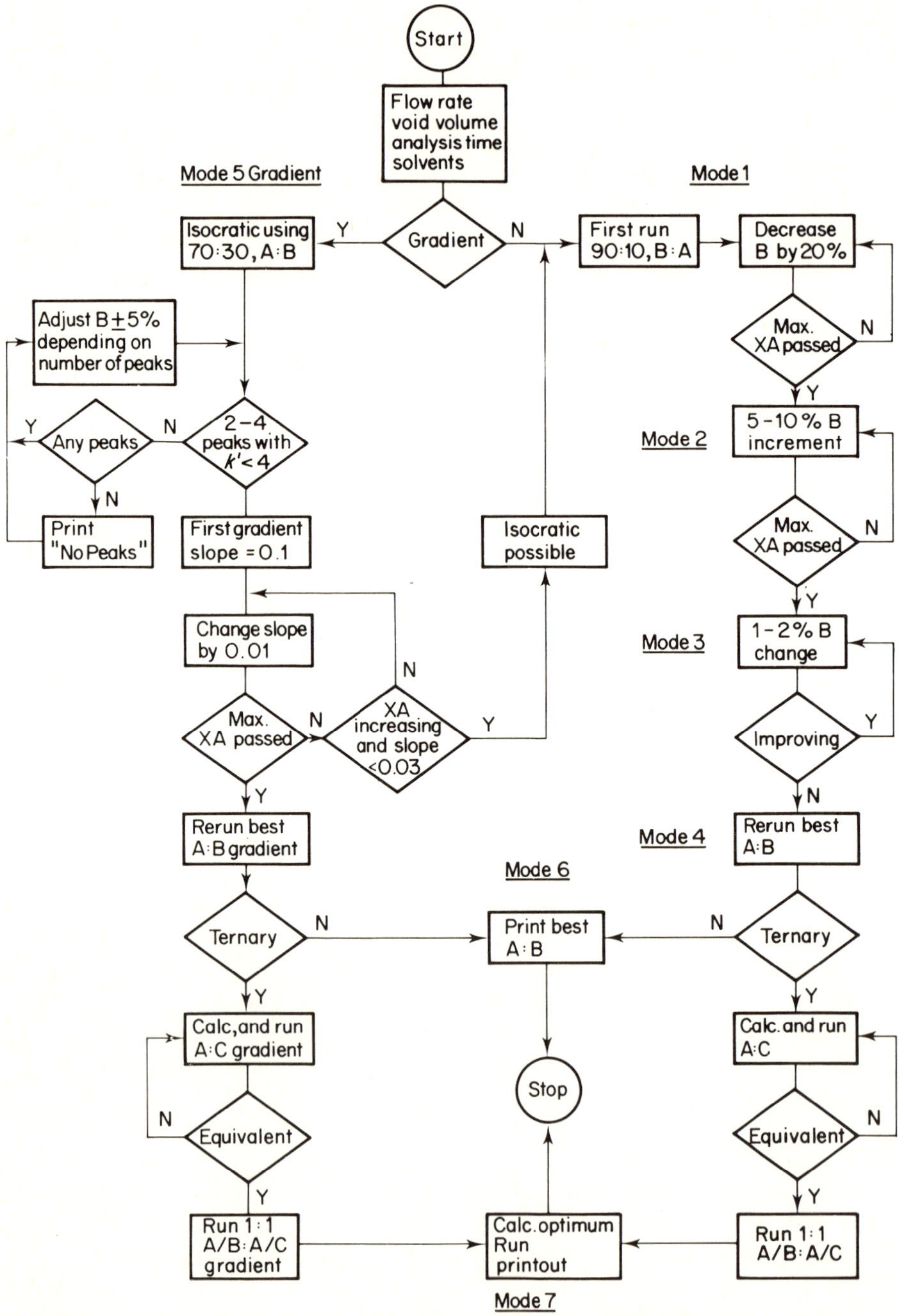

Figure 6.1. Flow diagram of optimization algorithm employed by the OPTIM 1 system. (Adapted from refs 1, 2 and 4)

until the criterion is met. If less than two peaks are found then the amount of solvent B is increased until the criterion is met.

2. An initial gradient run is carried out starting from the composition determined in Step 1 with the amount of B increasing at 10%/min. If the analysis time is too long the slope of the gradient is increased, otherwise the slope is decreased until XA is maximized. All changes in gradient slope are made in 1%/min. increments.
3. Where a ternary gradient is desired, the optimization procedure continues in a manner entirely analogous to that used for isocratic optimization. The gradient separation is rerun under the best A–B conditions and the equivalent A–C conditions are then calculated and run using the same value for the gradient slope. If the results are not approximately equivalent the conditions are adjusted until near equivalence is obtained. This equivalence is determined by comparing the average weighted retention times of all the peaks. A third separation is run using a 1:1 mixture of A–B and A–C. From the results of these separations the optimum conditions are calculated.

Should the gradient optimization procedure result in a gradient that is essentially flat, the software will recognize this and default to an isocratic optimization. The other default that will occur is a reversion to a binary separation if there is no obvious advantage to be gained from the use of a ternary solvent mixture.

6.1.2. Applications

For a procedure that appears to offer a relatively uncomplicated approach to separation optimization, it seems strange that there are few published applications. The capabilities of the isocratic procedure were demonstrated in the optimization of the reversed-phase separation of a 12-component synthetic mixture[1] using water, methanol and acetonitrile as the mobile phase components. The time taken to reach the optimum composition was 12 hours.[4] The gradient optimization capability of this method was used[2] in the optimization of the reversed-phase separation of phenolic acids in wine. In this example a ternary gradient was permitted but the final gradient recommendation was for water/acetonitrile.

6.1.3. Software

There are no published listings describing the 'adaptive intelligence' approach to separation optimization. As marketed by Spectra-Physics, the OPTIM 1 software is provided on an Erasable Programmable Read-only Memory (EPROM) chip designed to be plugged into one of their computing integrators. However, as can be seen from the flow diagram presented, it should be a relatively straightforward task to convert the algorithm as presented to a program for a suitable chromatographic system. In this case one will also be able to take advantage of experimenting with one or more of the chromatographic response functions that have been described in Chapter 2.

Another advantage that is obviously available to those writing their own software is the ability to modify the optimization algorithm. In the gradient optimization sequence a restriction is implicit at the first experiment (an isocratic run using a 70:30 A–B mixture). The software examines the first chromatogram for the appearance of peaks with k' less than 4. The decision-making process assumes that if fewer than two peaks are found eluting with k' less than 4 then the eluting power of the solvent is too low. The solvent strength is therefore increased by the addition of a further 5% of solvent B. While this reasoning may be true in many cases, it will be false if the components happen to be very poorly retained on the chosen column and have all been eluted at the solvent front. To cover all eventualities it is probably advisable to start with a much lower proportion of the first organic modifier, possibly none at all. Maybe the user could be given the choice of selecting the initial starting composition for the first isocratic experiment to cover the possibility that some knowledge of the sample is available? Indeed, some of these restrictions have been recognized and accounted for in the revised version OPTIM 2.

6.1.4. Summary

The 'adaptive intelligence' approach is an example of a method which uses information gathered during the optimization procedure to direct and modify the course of the optimization. Each chromatogram is evaluated and a decision taken as to the conditions required for the next experiment. The decision-making process is relatively simple, which obviously reduces the complexity of the controlling software.

Advantages of 'adaptive intelligence'

1. No prior knowledge of the number of components is required.
2. Applicable to both isocratic and gradient separations.
3. Relatively easy to program and implement on suitable hardware.
4. Comparatively fast, typically 10–20 hours.

Against

1. A 'local' rather than global optimum may be located.
2. Restricted to ternary reversed-phase or binary normal-phase separations, currently not applicable to ion-pair separations, etc.
3. In isocratic modes, all peaks must elute with 90% organic modifier.
4. For gradient separations all peaks must not be eluted with the solvent front using 30% organic modifier.

6.2. THE SEQUENTIAL SIMPLEX PROCEDURE

6.2.1. History

In 1951 a revolution occurred in statistical methods for experimental optimization, when the first unified discussion of the concepts of response surface methodology was published by Box and Wilson.[5] They further combined the response surface methodology with hill-climbing techniques to produce an algorithm capable of varying many factors simultaneously to discover the optimum level of response. At that time the technique was known as EVolutionary OPeration (EVOP).[6] EVOP essentially combined factorial designs with regression techniques to estimate the direction of steepest ascent and to locate the optimum region of a response surface. The great drawbacks of the original EVOP methods were the large numbers of experiments often required and the complex calculations that were necessary. In 1962 a new method was introduced by Spendley *et al.*[7] that overcame these major limitations. The method '. . . sprang originally from consideration of how Evolutionary Operation might be made automatic—a possibility to which Box drew attention even in 1955. More specifically, two problems were posed, viz. (i) could an Evolutionary Operation procedure be devised which would more rapidly approach and attain optimum conditions, and (ii) could the calculations and decisions be so formalized and simplified that they could be executed automatically by a digital computer . . .': the sequential simplex procedure was born. It can be seen from this quotation that right at the outset, the simplex procedure was aimed at being used for automated computer control of instruments and experiments.

The simplex method is guided by calculations and decisions that are rigorously specified yet almost trivially simple. The procedure is a hill-climbing method whose direction of advance is dependent solely on the ranking of responses and not on any particular values on an absolute scale. Two great features of the simplex method are that a new factor, previously held constant, can be introduced at any stage simply by conducting one extra experiment and that only one experiment (occasionally two, depending on the precise method being used) is needed for each step of the optimization. So many optimization methods require a new start if an additional parameter is to be included and little or no information is available on the whereabouts or direction of the optimum until all experiments have been conducted. The method is therefore highly efficient, multi-factor and has an empirical feedback allowing rapid attainment of the experimental optimum.

Note that the sequential simplex procedure is conceptually different from the simplex lattice design methods that were described in Chapter 4 for the overlapping resolution mapping method.

The sequential simplex procedure is a general optimization method and provides another example of the increasing number of chemometric procedures[8,9] being used in chromatographic optimization. The basic method of Spendley *et al.*,[7] 'the sequential application of simplex designs in optimization and evolutionary operations', will be described first. As will be seen, this original method has a number of limitations which were overcome by the powerful, yet simple, modification of Nelder and Mead.[10] It is the modified simplex procedure that is most often used today and the method which has, almost without exception, been applied to chromatographic optimization. Although a number of variants of the modified procedure itself are now available, these will not be described in great detail.

6.2.2. The Simplex Algorithm

A simplex is a geometric figure described by a number of points (vertices) equal to one more than the number of factors, or variables, being optimized. A simplex for two factors is a triangle. A simplex for three factors is a tetrahedron. The series can obviously be extended to higher dimensions but then the simplexes are not easily visualized. The complexities of the mathematics do not, however, increase significantly as the number of dimensions increases and the procedure is easily handled by computer for multi-factor optimization. In Figure 6.2 a

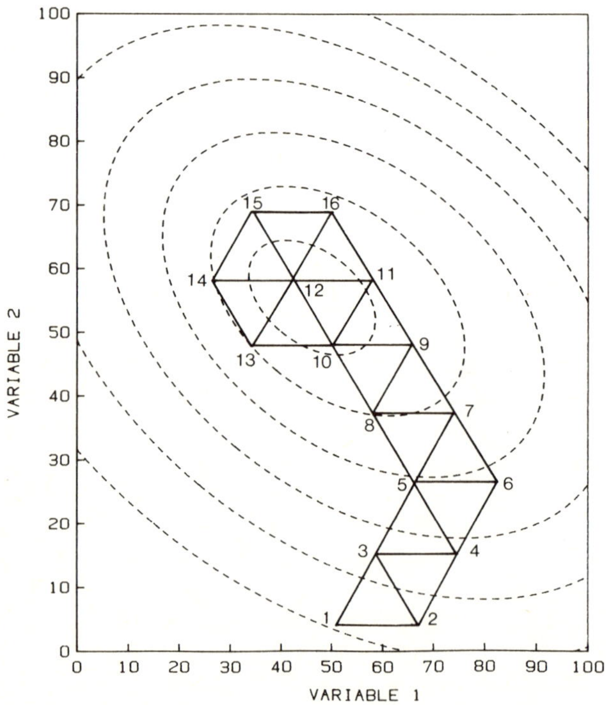

Figure 6.2. Movements of fixed step size simplex on response surface. The initial simplex is 1, 2, 3 and the optimum region lies close to point 12

two-dimensional simplex, that is a simplex arising from the optimization of two variables, is shown superimposed on a response surface contour map. The contour lines represent lines of equal response as a function of the two variables and the optimum is located at approximately variable 1 = 50, variable 2 = 55. In chromatographic terms the two variables might be the proportions of two solvents and the response might be the result of the calculation of a chromatography response function.

The objective of the sequential simplex procedure is to force the simplexes to move away from regions of poor response towards the region of optimum response. In Figure 6.2, a single experiment is carried out at the points 1, 2 and 3 to determine the experimental response at each of these points. The next and subsequent steps in the method are made according to a set of 'rules'.

RULE 1. A move away from the worst response is made after each experiment
Having carried out an experiment at each of the points 1, 2 and 3 in the starting simplex, the experimental responses are classified and ranked in the order best (B), next to worst (N) and worst (W). A new simplex is then constructed adjacent to the old simplex by applying Rule 2: only one additional experiment is required to complete each new simplex.

RULE 2. A move is made into the adjacent simplex which is obtained by rejecting the point with the worst response and replacing it with a point obtained by a reflection through the midpoint of the remaining (hyper)face
Figure 6.2 shows the progress of a two variable simplex towards an optimum. Point 1 of the original simplex had the lowest response (W) and is discarded, leaving points 2 and 3. Reflection of 1 through the midpoint of the line 2–3 generates the point 4. The new simplex is then formed from 2, 3 and 4. Repeating the process with this second simplex leads to point 2 being rejected and a new simplex 3, 4, 5 is then generated. This procedure is repeated until the optimum is located or until a violation of Rule 3, 4 or 5 occurs.

RULE 3. If the new point has the worst response in the new simplex, do not re-apply Rule 2 but instead reject the next to worst point in the new simplex
If this rule is not applied a sudden failure of the procedure may occur as the next simplex generated will be the same as the one just abandoned. If this were to happen the simplexes would end up oscillating across a hyperface—this kind of failure would be most likely on a response surface ridge. Referring to Figure 6.2, the reflection of point 11 gives a new vertex at point 13. Unfortunately, point 13 is the worst vertex in the new simplex 10, 12, 13 and so would be reflected back to give point 11 again. The correct application of Rule 3 means that point 10 is reflected to give the new simplex 12, 13, 14.

Note that so far all attention has been paid to identifying and dealing with the experiment that gives the worst response. In the real world all experimentation is subject to errors and to noise and we need to be sure that the best experimental point is not falsely high and therefore misleading the progress of the procedure. It is possible that a falsely high result will lead to the simplexes becoming attached to that point and therefore prevented from moving across the response surface. To aid in distinguishing between a true optimum and a falsely high result, the following exception to Rule 1 is used.

RULE 4. If a vertex has been retained in n + *1 simplexes, where* n *is the number of variables, before re-applying Rule 2 re-evaulate the response at that vertex*
This rule prevents the simplex from becoming stranded on an experimental point whose result is falsely high. If the vertex is truly near the optimum, it is likely that the repeated result will also be high and the maximum will be retained. If the original result was in error then it is unlikely that the repeat observation will also be in error and that vertex will eventualy be eliminated. Note that the simplex method is fairly tolerant of mistakes and a move in the wrong direction will be quickly corrected by Rules 2 and 3.

The final rule is designed to take care of the situation where the procedure suggests that an experiment should be conducted in a region outside the normal experimental boundaries.

RULE 5. If a new point lies outside the boundaries of the independent variables do not make an experimental observation but instead assign it a very undesirable response
It is obviously necessary to constrain the optimization within some reasonable boundaries or factor space. Recognizing that situations will arise when the procedure attempts to suggest an experiment outside this factor space, the assignment of a very bad result ensures that this point will have the worst result in the present simplex. This very poor response is not ignored but is included in the application of Rules 2 and 3 which will then force the simplex back inside the boundaries and the procedure will continue to seek the optimum response.

When the optimum has been located, the rules force the simplexes to circle around the optimum. As can be seen from Figure 6.2, this circling will probably encompass the optimum but is unlikely to locate it precisely. In the example shown it is likely that the optimum is positioned somewhere between points 10 and 12. It is wise in the early stages of optimization to choose a fairly large initial simplex to locate the general area of optimum response and then to start a new, smaller simplex in that region to define the optimum more precisely. This optimum may be located even more exactly by using a still smaller simplex, except that there is a practical limit to the size of the simplexes. If the simplex size is too small, the differences in responses will be masked by the indeterminate errors. In such a situation the simplexes will wander erratically in the region of the optimum.

There is always the possibility that this located optimum is not the true, global, optimum but a local optimum. There is no way of being absolutely certain that a simplex optimization has located the overall optimum. Consider as an example the separation of three components, A, B and C. One optimized separation may result in the elution order A, B, C whilst another may yield A, C, B. The overall optimum separation may require that the elution order is C, A, B. Confidence in the quality and nature of the optimum can be increased if the same optimum is located when the procedure is started from very different parts of the available factor space.

The basic, fixed step size simplex procedure of Spendley *et al.*[7] suffers from at least three limitations in normal use. The first problem is that, especially for more than two variables, it is not always easy to be sure an optimum has been located. In two dimensions (a triangular simplex) the simplexes will circle and will be superimposable because of the ability of triangles to close pack (see Figure 6.2). This is not necessarily the case in higher dimensions as tetrahedra and higher dimensional simplexes will not close pack and it becomes less easy to determine that an optimum has been located. Secondly, there is no provision for changing the size of the simplex to accommodate the changing nature of the response surface. This problem is generally overcome by running several different optimizations with progressively smaller simplexes. Finally, the fixed step size simplex can still suffer failure on a ridge, despite the application of Rules 3 to 5. If the step size is too large and the orientation of the simplexes just right, the simplexes may straddle a ridge without any indication that this is a false optimum. The only way of checking this is to alter the orientation of the simplexes slightly and repeat the optimization. Thus there are features of the fixed step size simplex which detract from its general appeal. Nevertheless, many applications are ideally suited to the use of the fixed step size simplex. For example, there are situations where the optimum is not stationary and a simplex procedure is required first to locate that optimum and then follow it across a response surface that is changing with time. This is a recurrent problem in analytical chemistry in relation to keeping instruments under their best operating conditions as the performance of their various components, for example the chromatographic column, changes with age.

6.2.3. The Modified Simplex Procedure

The modifications by Nelder and Mead[10] to the original simplex method largely overcome all the limitations to the basic method. A much clearer indication of the attainment of a stationary optimum is provided and the chances of failure on a ridge are also reduced. The greatest advantages are in the ability of the method to accelerate the progress of the simplexes and to adapt the simplex size in accordance with the response surface. In addition to the operation of *reflection*, two new operations are introduced—*expansion* and *contraction*. Together these operations provide the means for the simplex figures to expand and accelerate towards the optimum region and, having approximately located this region, to contract and reduce the search region until the optimum is precisely located.

The movements of the simplexes are still governed by the same basic rules that were used for the fixed step size procedure. Additional tests are included to decide which other operation to perform in a particular situation. Consider

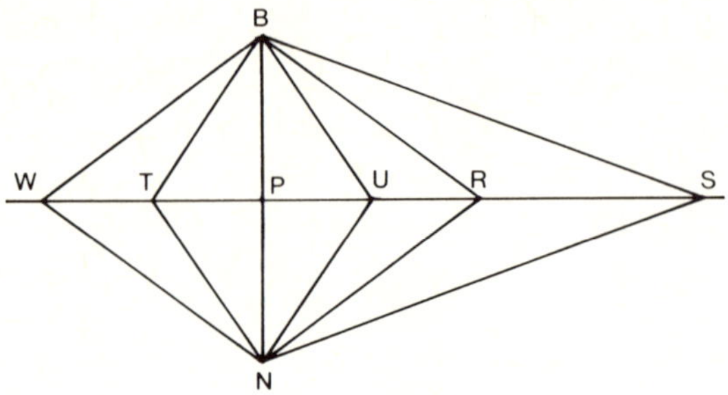

Figure 6.3. The modified simplex procedure: the initial simplex is B(est), N(ext to best) and W(orst). Reflection of W through P gives R. The subsequent move could be expansion (to S), a contraction close to R (U) or a contraction towards W(T)

the initial simplex BNW shown in Figure 6.3. B is the vertex with the best response, W has the worst response and N is the next to worst vertex. P is the centroid of the hyperface BN which remains when the worst response W is rejected. Reflection of the point W through P generates the next point to be investigated R. In vector notation:

$$\mathbf{R} = \mathbf{P} + (\mathbf{P} - \mathbf{W}) \tag{6.1}$$

There are three possibilities for the measured response at R.

i. The response at R is even better than the response at B, which suggests that the simplex is moving in the right direction and that further investigation of this *direction* should be carried out. Expansion of the segment WR is therefore indicated and a new simplex BNS is generated according to

$$\mathbf{S} = \mathbf{P} + a(\mathbf{P} - \mathbf{W}) \tag{6.2}$$

where a is the expansion coefficient, usually set to 2, which determines how far the segment is to be expanded. If S turns out indeed to be better than B, then this new simplex BNS is retained. If the response at S is not better than B, then the new simplex is taken as BNR and the expansion is said to have failed. Note that in some modifications of the procedure the response at S is compared not with B but with R.[11,12]

ii. The response at R is neither better than the response at B nor worse than the response at N. In this somewhat ambiguous situation neither expansion nor contraction is clearly indicated and the new simplex is taken as BNR.

iii. The response at R is worse than the response at N. This clearly indicates that the simplex is trying to move in the wrong direction and suggests contraction and a movement away from this region. If the response is even worse than that at W, then the direction is wholly unsatisfactory and the new, contracted point should lie at T, which is closer to W than R:

$$\mathbf{T} = \mathbf{P} - b(\mathbf{P} - \mathbf{W}) \tag{6.3}$$

where b is the contraction coefficient, usually set to 0.5.

If the response at R lies between the responses at W and B, then the new, contracted point should be at U, which is closer to R than W:

$$\mathbf{U} = \mathbf{P} + b(\mathbf{P} - \mathbf{W}) \tag{6.4}$$

If T or U subsequently gives the worst result in the new simplex, there is a possibility that the next steps may result in the loss of movement along the W–R direction. This results from the worst vertex collapsing on to the point P—the original triangle has now become a straight line. One degree of freedom will be lost and the simplex may become stranded: this is the problem of 'degeneracy'. Degeneracy is not an inevitable fate of the simplex procedure, as corrective action can be taken to ensure that the simplex does not become 'stuck'. Such courses of action include a reduction of the total simplex size[10] or a translation of the simplex,[13] that is a change in the orientation of the

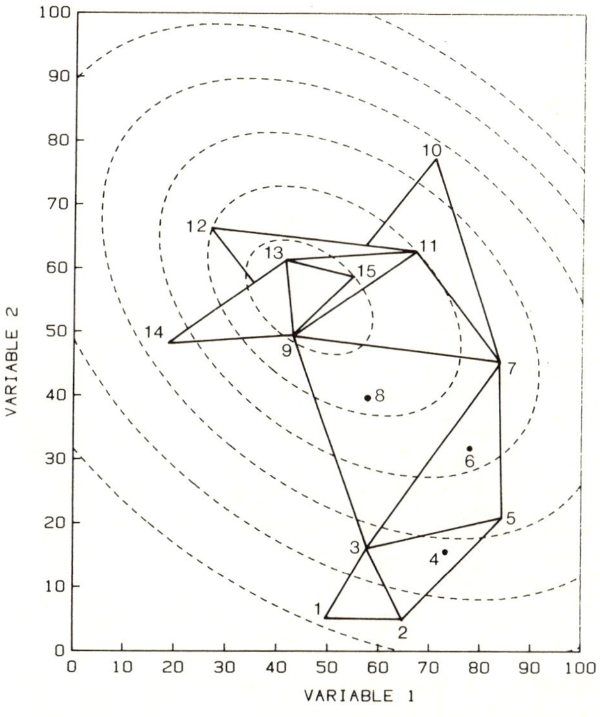

Figure 6.4. Movements of the modified simplex procedure across the response surface as used in Figure 6.2. The initial simplex is 1, 2, 3 and the final simplex is 9, 13, 15

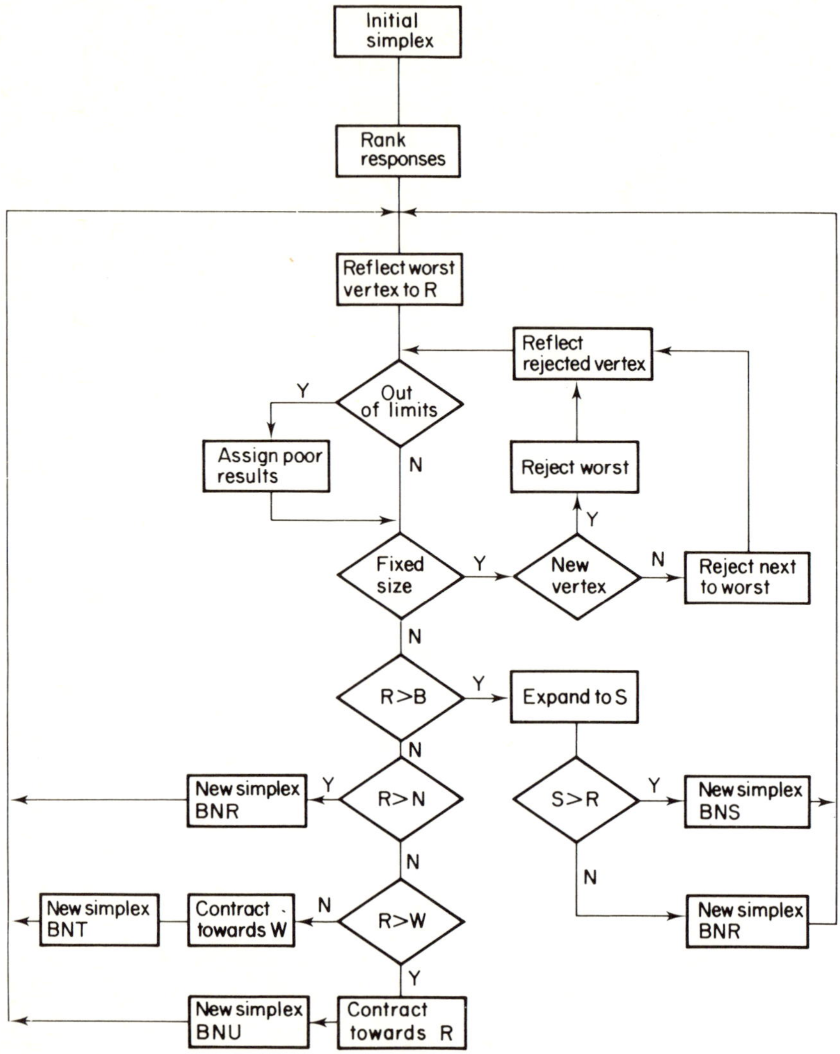

Figure 6.5. Flow diagram for fixed and modified simplex procedures

vertices or a complete transportation to another region of the experimental variables.

Figure 6.4 shows the progress of a variable step size simplex on the same response surface shown for the fixed step size simplex in Figure 6.2. It is evident that the optimum region is approached more rapidly and that the simplex size contracts to more precisely locate the true optimum. Figure 6.5 summarizes the rules of the simplex procedure for vertex acceptance or rejection for both the fixed step size and modified simplex procedures.

There have been many developments of the modified simplex procedure, all intended to improve speed and/or reliability. While the modified simplex procedure utilizes fixed values of the expansion and contraction coefficients (usually 2 and 0.5), the super-modified simplex (SMS) method[14] determines optimum values for the coefficients according to a parabolic fit through the worst vertex, the reflected vertex and the mid-point.

Both the modified and super-modified procedures reflect the worst vertex through the mid-point of the remaining hyperface but, referring again to Figure 6.3, one might anticipate that the optimum lies more in the region of B rather than N. The Weighted Centroid Method (WCM)[15] attempts to take this into account. Instead of a reflection through P, the reflection will be through a point closer to B (P′ in Figure 6.6). The use of the WCM is reported to speed up the attainment of an optimum but it does suffer more than the modified simplex method in the presence of noise. A further problem which can be experienced with the WCM is degeneracy, which tends to lead more quickly to simplexes which are distorted. As all reflections and expansions are deliberately biased towards the vertex giving the best response, the ability to search in directions perpendicular to the previous search direction is diminished.

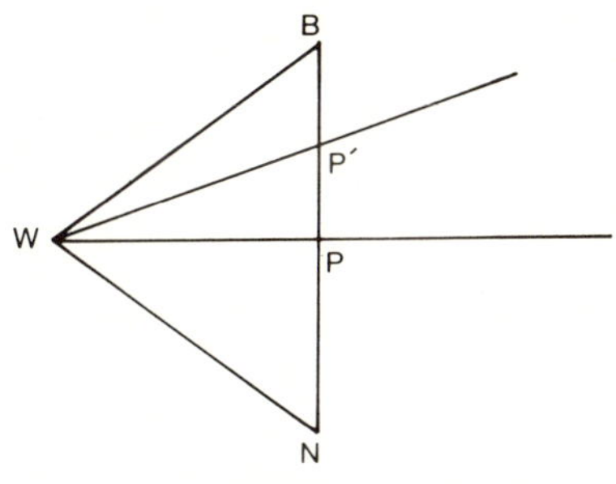

Figure 6.6. The weighted centroid simplex procedure: the direction of reflection lies not through P, as would be the case in the standard simplex procedures, but through P′, which is situated closer to B(est)

One way of reducing the problems of degeneracy is to introduce a symmetry criterion.[16] A perfect regular simplex is defined as having a symmetry of 1 with a degenerate simplex having a symmetry of 0. Provided the symmetry exceeds a minimum value, the WCM can be used. If the symmetry threatens to fall below the threshold, then an SMS method is reverted to.

Ultimately, it will be necessary to decide which variant of the simplex procedure will be used. There are small, but significant, differences in the abilities of the different methods to cope with various challenges such as noise and drift. In chromatographic optimization it is likely that the data will have a degree of superimposed noise. In such situations the modified simplex procedure has been shown[17] to out-perform the weighted centroid method, particularly when the expansion vertex is compared with the reflection vertex (i.e. point S is compared with point R rather than point B in Figure 6.3). Additionally, the calculations associated with the modified procedure are relatively easy and it is the variant which is currently most frequently used for chromatographic optimization.

The two aspects remaining to be discussed for simplex optimization are the stopping and starting of the method. The halting of the simplex movements is relatively simple and can be based on a minimum step size (e.g. 1% of the span of each variable), the number of experiments carried out or when the differences in responses between vertices approach the random noise and error. Some workers[16] have used the determinant of the simplex as a measure of its size and have used the determinant as a stop criterion. If the number of components in the sample and the required resolution are known, a chromatography response function can be employed which reflects this knowledge.[18] The optimization procedure can be halted as soon as the function meets some pre-specified value.

The location of the initial simplex and the decision as to its starting size are more difficult. The size of the initial simplex is important. A small simplex may give a more detailed evaluation of the response surface but the differences in response at each vertex may be too close to the random noise. A large starting simplex should ensure that substantial response differences are seen. An additional benefit is that a large simplex will be forced to reflect back inside itself, thereby giving an outline of the chromatographic behaviour within just a few experiments. The choice of the step size is arbitrary but it is highly desirable that a step size for each factor is chosen which will ensure some change in response. Ideally, the change in response should be comparable for a similar sized step for all the variables under consideration (to avoid the degeneracy problem described above) but this is, unfortunately, seldom predictable. An initial large step size is an advantage for two reasons. Firstly, the change in response is almost certain to considerably exceed the random noise of the system and secondly, a larger step can more rapidly approach the optimum. Where the modified simplex procedure is being used, the simplexes may continue to expand as the optimum is approached and the problem of overshoot arises. Contraction and reflection should overcome this problem but a significant number of experiments will have to be conducted before the simplex size has contracted sufficiently to locate the optimum precisely.

A decision also needs to be made as to the siting of the first vertex; this decision is strongly related to the simplex size selection. It is often considered good practice to have this first vertex at a point where at least some response is expected. For the automated chromatographic optimization of an unknown mixture it is likely that little knowledge will be available to aid this decision. If the determination of solvent composition is the target of the optimization, then a reasonable point for the first vertex might be close to, or even at, the upper limit of the most powerful eluting component under investigation; at this point there is the reassurance that all solutes will be eluted. The positions of the other vertices can then be established automatically by the application of the general rules described by Spendley *et al.*[7] Consider a two-dimensional simplex with the first vertex at coordinates 10, 10 and each variable having the range 0–100. The other two vertices will be located at $(10+a, 10+b)$ and $(10+b, 10+a)$, where

$$a=\frac{S_n}{n.2^{1/2}}[(n+1)^{1/2}+(n-1)]$$

$$b=\frac{S_n}{n.2^{1/2}}[(n+1)^{1/2}-1]$$

where n is the number of variables and S_n is the step size for each factor. If the step size is set at 0.6, the other two vertices will then be located at (26, 68) and (68, 26). Where more than two variables are being considered, the coordinates of the initial simplex can be calculated from Table 6.1. Further guidance on the location of the initial simplex can be found in a review by Yarbro and Deming.[19]

TABLE 6.1. Coordinates of initial simplex (after ref. 19)

	Dimensions					
Vertex	1	2	3	4	$n-1$	n
1	0	0	0	0	0	0
2	a*	b†	b	b	b	b
3	b	a	b	b	b	b
4	b	b	a	b	b	b
5	b	b	b	a	b	b
n	b	b	b	b	a	b
$n+1$	b	b	b	b	b	a

$${}^*a=\frac{S_n}{n.2^{1/2}}[(n+1)^{1/2}+(n-1)].$$

$${}^\dagger b=\frac{S_n}{n.2^{1/2}}[(n+1)^{1/2}-1].$$

6.2.4. Applications

Being a general optimization technique, the sequential simplex procedure has found successful applications in an enormous variety of fields.[20-25] The simplex procedure was relatively late in appearing in chromatographic optimization, it not being until 1975 that examples in both gas[26] and liquid[27] chromatography appeared. The liquid chromatography application was a pioneering experiment to test the applicability of the simplex method to chromatography using the ion-exchange separation of five inorganic cations as the experimental system. The potential of the method was clearly proved but, even at this relatively early stage, it was realized that a most important difficulty to be overcome was the choice of the response factor to be optimized. This is, of course, fundamental to the successful operation of the method: there must be a reliable and meaningful criterion to be optimized. Two years later, the first HPLC application was published,[28] describing the optimization of the solvent composition for the normal-phase separation of phospholipids. Once again the problems of assessing chromatographic performance were recognized. Additionally, where three variables were being optimized, it was found difficult to establish the precise direction in which the simplexes were moving and whether an optimum had actually been located.

Recognizing the importance of the optimization criterion, Watson and Carr[29] developed a chromatographic response function (CRF) to use with the modified simplex procedure for the gradient optimization of the reversed-phase separation of five PTH-amino acids. The CRF used was

$$\mathrm{CRF} = \Sigma \ln(P_i/P_0) + a(T_{\mathrm{M}} - T_{\mathrm{L}}) \quad (6.5)$$

where P_0 and T_{M} are the desired peak separation and total analysis time and P_i and T_{L} are the achieved resolution and analysis time, respectively, and a is an arbitrary weighting factor. Initial experiments examined just two factors, namely mobile phase flow rate and isocratic composition for the optimization of three of the components. Good separations were achieved in 11–18 experiments, but it was shown that there was little to be gained from the systematic optimization of the flow rate in such a simple system. The next stage was to consider a more complex optimization of five variables: a weak and strong solvent, the gradient shape, the gradient duration and the mobile phase flow rate. In this case 20 experiments were required to produce an optimum separation.

Svoboda[30] examined the reversed-phase separation of 12 nucleotides using both a systematic (grid search) and simplex search procedure. The eluent had three variables: organic modifier content, buffer concentration and pH. Additionally, separation time and column length were also considered. It was concluded that the simplex search was better suited to the more complex situations, such as dealing with a multicomponent eluent, but that simple situations, such as a single-component eluent, were better dealt with by systematic searching.

Gradient optimization was also the objective of Fast *et al.*,[31] who used the CRF described by Watson and Carr (equation 6.5). Three factors were varied simultaneously to maximize the CRF: flow rate, column temperature and gradient shape. A total of 17 experiments was required to optimize the reversed-phase separation of five steroids. Ultimately, three simplex searches were carried out, each locating a slightly different set of optimum conditions, all of which were acceptable. This is the classic problem associated with simplex optimization: the optimum located may not be the global optimum but instead may be a lesser, local optimum. Nevertheless, given that it is an acceptable set of conditions, this does not detract from the appeal and usefulness of the method.

The first example of the fully automated optimization of HPLC separations under the guidance of the simplex algorithm was published in 1982 by

Berridge.[32] A new CRF was developed to give greater flexibility and to accommodate the needs of automated optimization:

$$\mathrm{CRF} = \sum_i R_i + L^a - b|T_\mathrm{M} - T_\mathrm{L}| - c(T_0 - T_1) \tag{6.6}$$

where R_i is the resolution between adjacent peak pairs, L is the number of peaks detected, T_M is the maximum desired analysis time, T_L is the retention time of the last eluted peak, T_0 is the minimum retention time for the first eluted peak and T_1 is the retention time of that first peak. Factors a, b and c are operator-selectable weightings (usually set to values between 0 and 3). These weightings can be varied according to the requirements of the problem in order to alter the importance of peak numbers, resolution or time with respect to the final separation.

Using a commercially available microcomputer-controlled chromatograph, three programs were developed to undertake three different aspects of reversed-phase chromatography:

ISOOPT	The optimization of an isocratic binary mobile phase composition and flow rate.
GRADOPT	The optimization of multi-linear gradient elution profiles.
TERNOPT	The optimization of isocratic ternary mobile phases.

All the programs use the modified simplex algorithm of Nelder and Mead[10] and a simplified flow diagram of the TERNOPT program is shown in Figure 6.7. Note that an arbitrary poor response of -100 is assigned to an experiment that tries to violate the boundary conditions. Should a chromatogram be produced which has no detectable peaks a response of -50 is assigned to that experiment. The program GRADOPT was shown to be able to optimize the gradient separation of a simple four-component mixture in 15 experiments, whereas ISOOPT and TERNOPT each required 29 experiments for (different) four-component mixtures.

It was subsequently shown that the simplex procedure can be used also for normal-phase chromatography[33] with the application of TERNOPT to the optimization of the separation of five aromatic solutes. This application is of particular interest since it includes an example of a method of reducing the chances that a local, rather than a global, optimum has been located. The overall strategy is illustrated in Figure 6.8. The key feature is that, at the conclusion of a given optimization run, the chromatogram at the optimum is subjected to double differentiation with respect to time. Differentiating the detector signal amplitude with respect to time gives a decrease in peak width[34] which can be used to great advantage to enhance the resolution between poorly resolved peaks. The results of this are shown in Figure 6.9, where it can be seen that the apparently optimized separation showing excellent resolution of four components is misleading since, in the second derivative, five components are

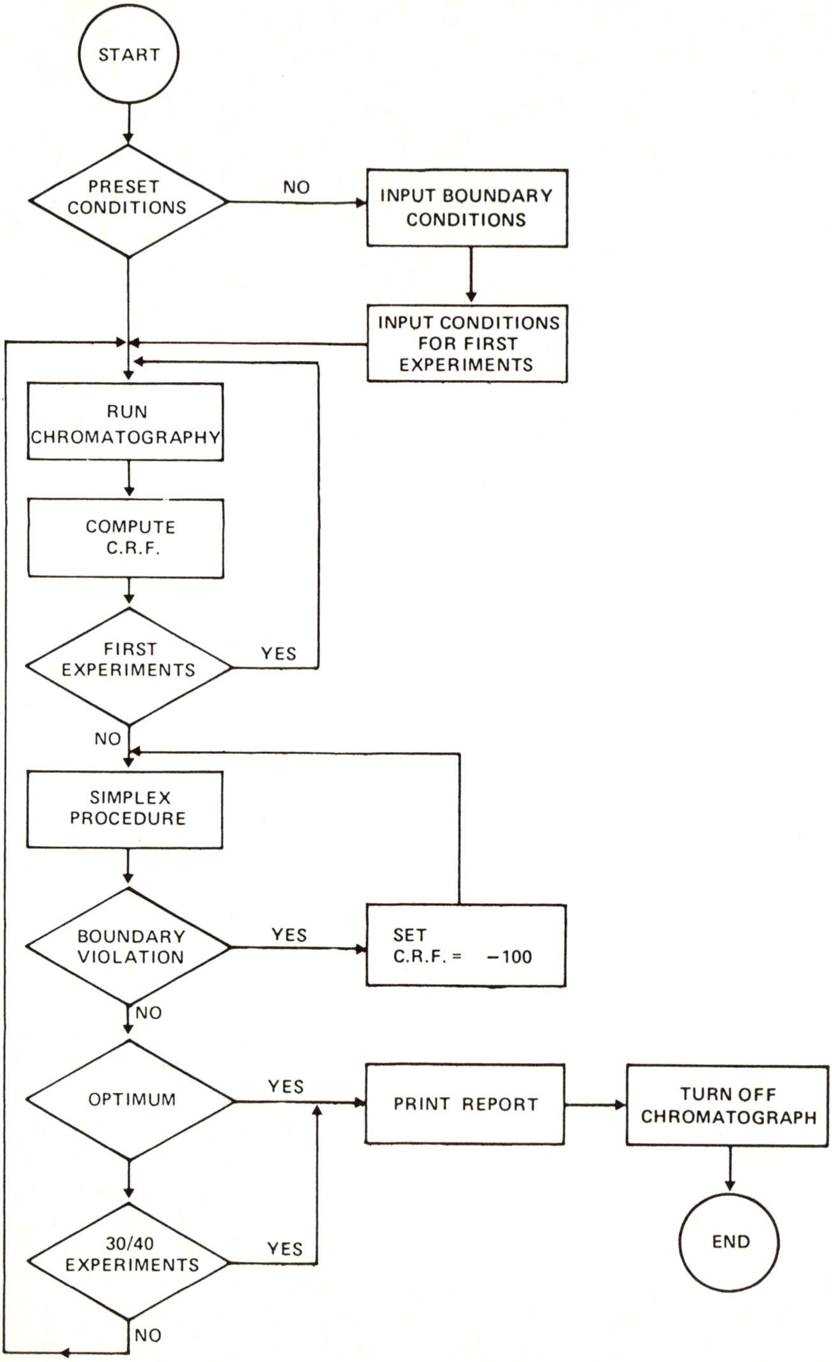

Figure 6.7. Flow diagram for simplex algorithm used in ISOOPT, GRADOPT and TERNOPT automated optimization programs

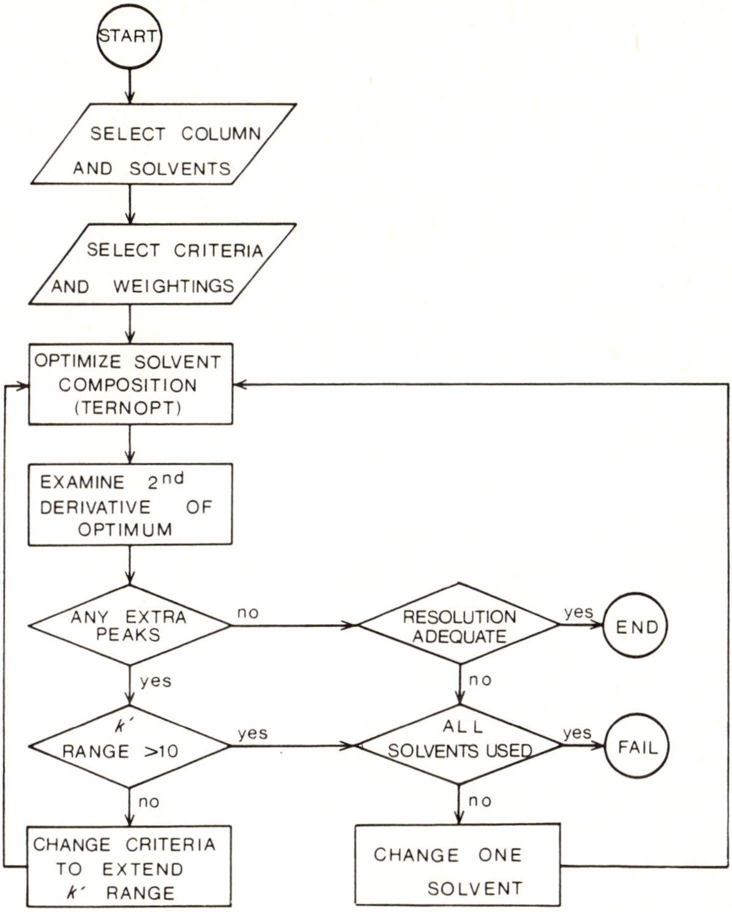

Figure 6.8. Flow diagram for extended optimization scheme which uses the second derivative to examine the quality of the achieved optimum. *Reproduced from ref. 33 by permission of Friedr. Vieweg & Sohn*

revealed. A second run with different time criteria and flow rate gave an improved but not acceptable separation (Figure 6.10). There is a clear separation of five components but full resolution has not been achieved. The second derivative produces essentially complete resolution of the peaks and no additional components are indicated. As the chromatographic system used was restricted to ternary mobile phases, the next practical step was to substitute one of the solvents and re-run the optimization. This third run provided the optimum separation, no further peaks being revealed from the second derivative (Figure 6.11). Whereas in normal-phase chromatography there is probably little to be

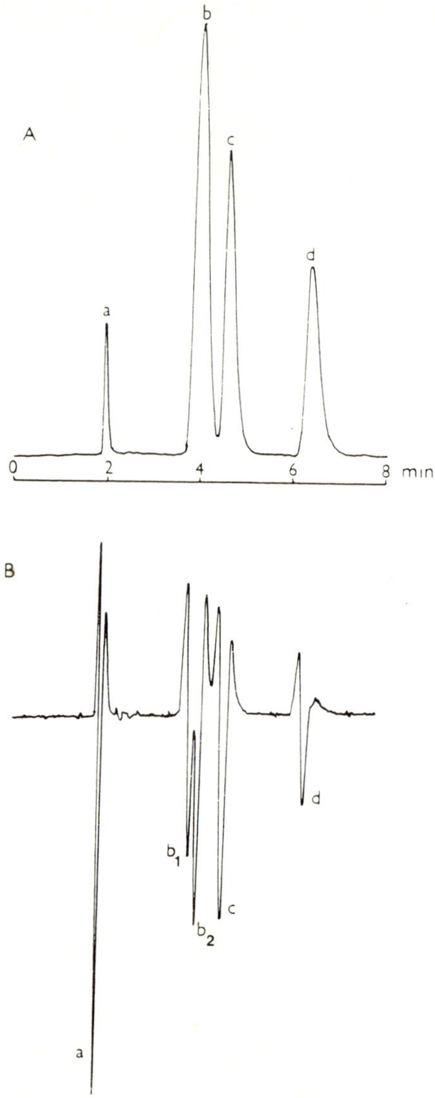

Figure 6.9. (A) Chromatogram obtained following first optimization using TERNOPT and (B) the second derivative. Column, 30 cm × 0.39 cm i.d. containing 10 μm micro-Porasil; mobile phase, hexane/ethyl acetate/methyl tert-butyl ether (77:8.5:14.5). Solutes: 1 = butylated hydroxytoluene; 2 = butyl paraben; 3 = propyl paraben; 4 = methyl paraben; 5 = benzyl alcohol. *Reproduced from ref. 33 by permission of Friedr. Vieweg & Sohn*

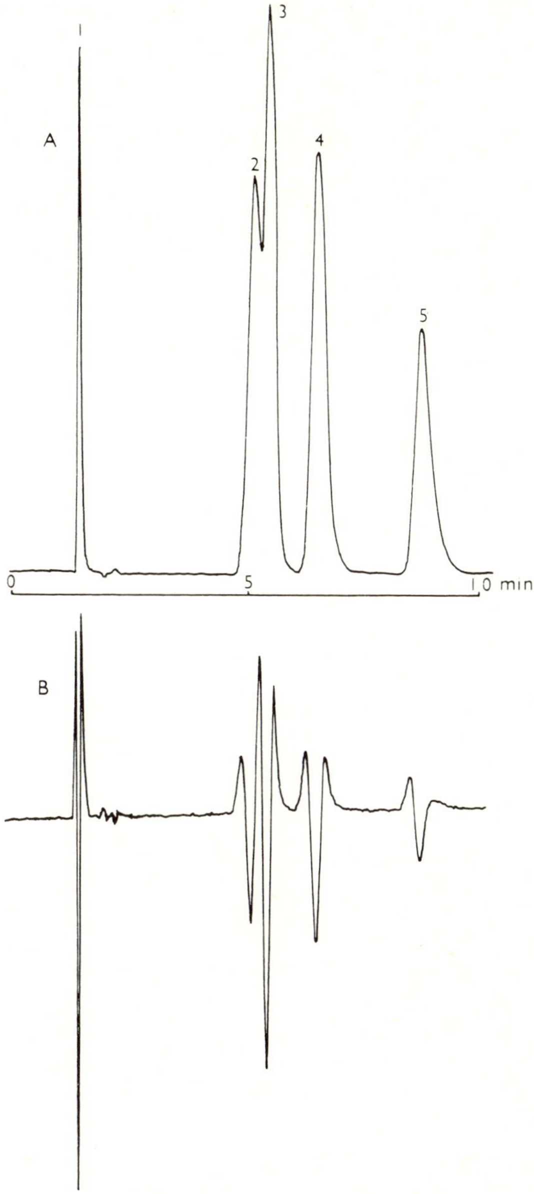

Figure 6.10. (A) Chromatogram and (B) second derivative of separation following second run of TERNOPT. Mobile phase, hexane/ethyl acetate/methyl tert-butyl ether (87:11:2). Solutes as in Figure 6.9. *Reproduced from ref. 33 by permission of Friedr. Vieweg & Sohn*

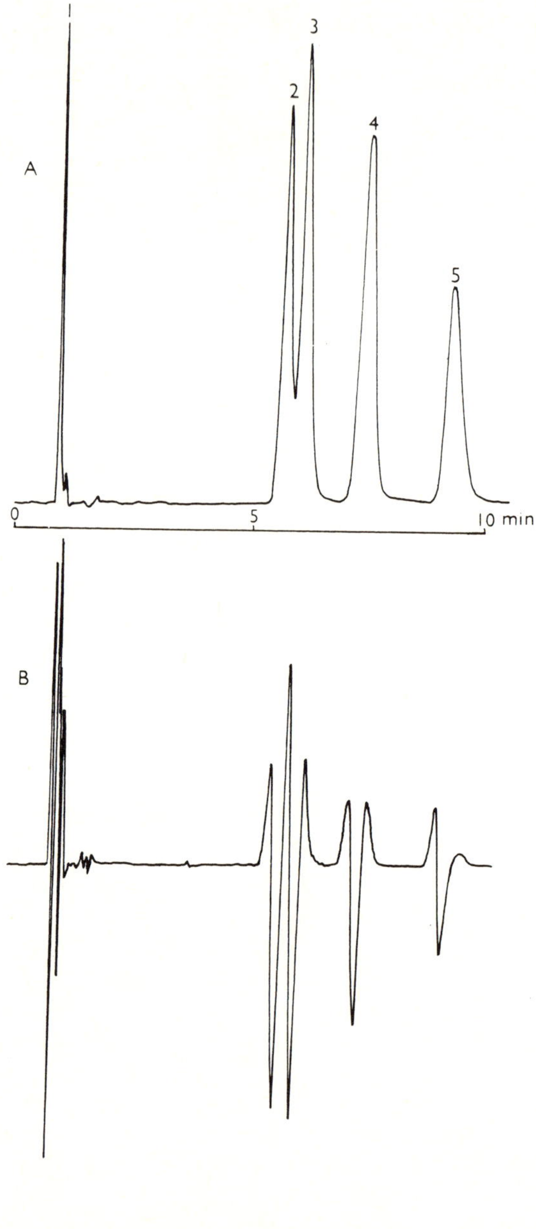

Figure 6.11(A). Chromatogram and (B) second derivative of separation following final run of TERNOPT after switching one mobile phase component. Mobile phase, hexane/ethyl acetate/methylene chloride (80:6:14). Solutes as in Figure 6.9. *Reproduced from ref. 33 by permission of Friedr. Vieweg & Sohn*

gained from substituting a different column should an optimum separation not be attained: this will not be the case in reversed-phase chromatography. The strategy could therefore be extended, if suitable chromatographic hardware is available, to include the selection of the column length or stationary phase.

The capabilities of the simplex procedure in totally automated separation optimization were extended with the program MULTOPT,[35] which includes ternary mobile phase composition, flow rate and temperature as the variables. MULTOPT has been used to optimize pH, ion-pair reagent composition, flow rate and temperature all simultaneously and automatically, requiring less than 40 experiments to locate optimum conditions for five- and six-component mixtures. One example described in this work is the optimization of the separation of six acids. This application is of particular interest since two of the three solvents employed (trichloroacetic acid and trichloroacetate buffer) are strongly interdependent, controlling both pH and ion-pairing aspects of the separation, showing the ability of the simplex procedure to cope with complex separation systems.

Two problems associated with the simplex optimization of separations of completely unknown mixtures are (i) that many of the early experiments may yield little information because they are conducted in regions which are very distant from the ultimate optimum and (ii) that local optima can often be found rather than the global optimum. The chances of these both happening can be considerably reduced by restricting the search area to a region of the factor space in which the global optimum must lie. Our knowledge of reversed-phase chromatography is such that this is indeed possible. By running an initial gradient separation of the mix it is possible to gain sufficient information about the mixture to reduce dramatically the area over which a search must be conducted. This is the principle behind another program, FASTOPT.[36]

The program FASTOPT (Figure 6.12) begins by conducting a reversed-phase gradient separation using methanol and water under conditions calculated to provide for maximum resolution with the system in use (see the discussion of gradient separations in Chapter 3). The elution times of the first and last detected peaks are examined and solvent strengths are calculated that will provide approximate translations to reasonable isocratic separations. Next, the proportions of other standard solvents for reversed-phase chromatography are calculated (see again Chapter 3 for methods of translating gradient results to isocratic separations). These results are used to define a reduced area with respect to solvent composition over which the simplex search is conducted. This restriction of the search area effectively reduces the time taken to locate a suitable solvent strength and gives a greater opportunity for the optimization of selectivity. The improvements in speed of attaining the optimum and in the reproducibility of that optimum are dramatic.

The separation of a mixture of five sulphonamides was used to illustrate the use of the method. This mixture was chosen to present a severe challenge to the procedure, since it had been suggested[37] that the optimization of its separation by sequential methods would be of questionable success. In a situation

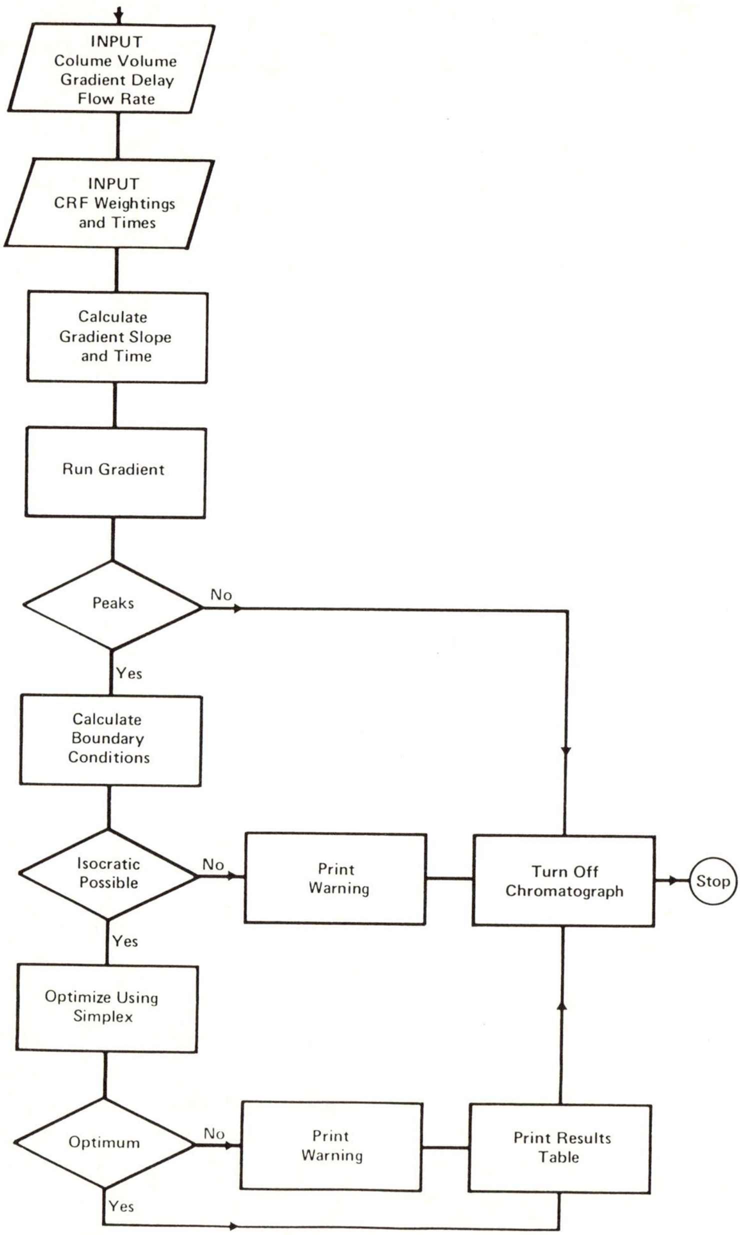

Figure 6.12. Flow diagram of FASTOPT program. *Reproduced from ref. 36 by permission of Elsevier Science Publishers BV*

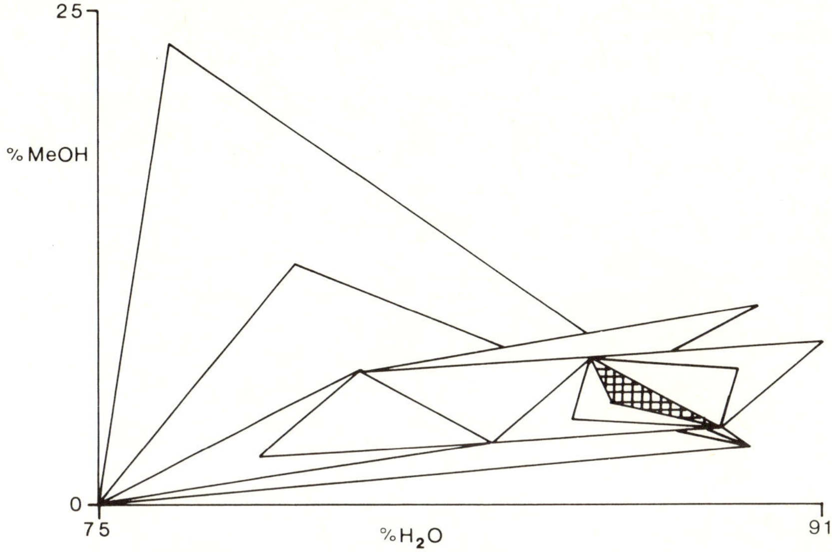

Figure 6.13. Movements of simplexes from the optimization of the separation of five sulphonamides under the control of FASTOPT. *Reproduced from ref. 36 with permission of Elsevier Science Publishers BV*

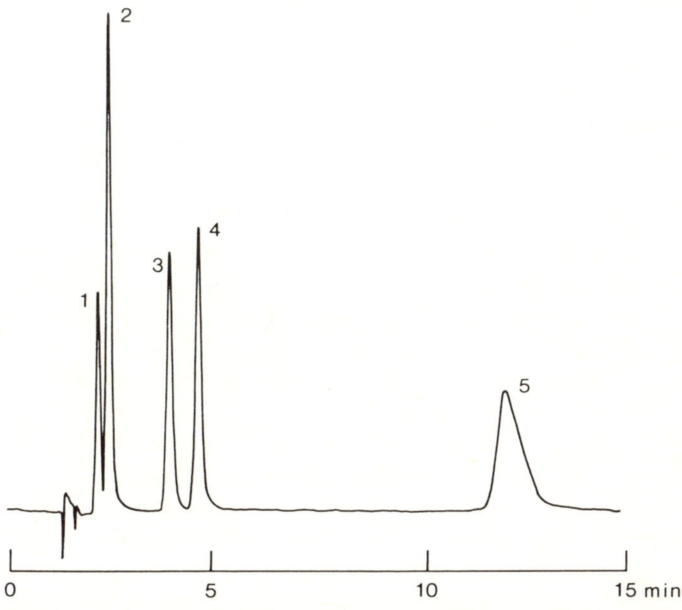

Figure 6.14. Optimum separation as located by FASTOPT for five sulphonamides. Column, 15 cm × 0.46 cm i.d. Ultrasphere ODS; mobile phase, water/methanol/acetonitrile (85:4:11). Solutes: 1 = sulphisomidine; 2 = sulphaguanidine; 3 = sulphanilamide; 4 = sulphacetamide; 5 = sulphadiazine. *Reproduced from ref. 36 with permission of Elsevier Science Publishers BV*

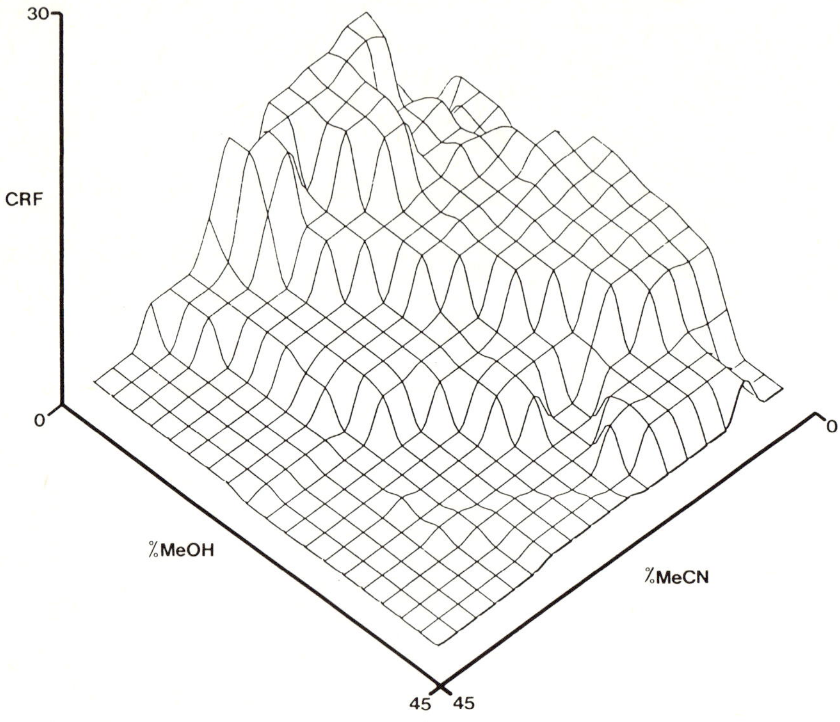

Figure 6.15. Response surface for the separation shown in Figure 6.14 using CRF as the quality criterion. *Reproduced from ref. 36 with permission of Elsevier Science Publishers BV*

where TERNOPT completely failed to locate an optimum separation, FASTOPT quickly and reproducibly located the global optimum even when started from completely different positions within the restricted search area. The initial gradient run enabled this restricted area to be limited to the range 75–91% of water. The two organic modifiers are thereby limited to the range 0–25%, bearing in mind that there must always be at least 9% of organic modifier present. The movements of the simplexes during one optimization run are shown in Figure 6.13, with the optimum solvent composition region being shaded. The separation achieved is shown in Figure 6.14. In Figure 6.15 the response surface of the separation is shown. This response surface was determined entirely experimentally, requiring about 400 experiments! The response surface does show that the CRF used (equation 6.6) generates a relatively smooth surface with a minimum of local optima. As each new peak is discovered the surface rises in a series of plateaux, each sloping gently upwards towards the optimum region.

TABLE 6.2. Instrument parameters and conditions

Column volume	0.5 ml
Gradient delay volume	1.5 ml
Flow rate	1.0 ml/min
Temperature	22°C
Detection wavelength	265 nm
Solvent A	Water
Solvent B	Methanol
Solvent C	Acetonitrile
Minimum retention time	1 min
CRF weighting	1
Maximum retention time	4 min
CRF weighting	1
CRF weighting for No. of peaks	2
Analysis time	6 min

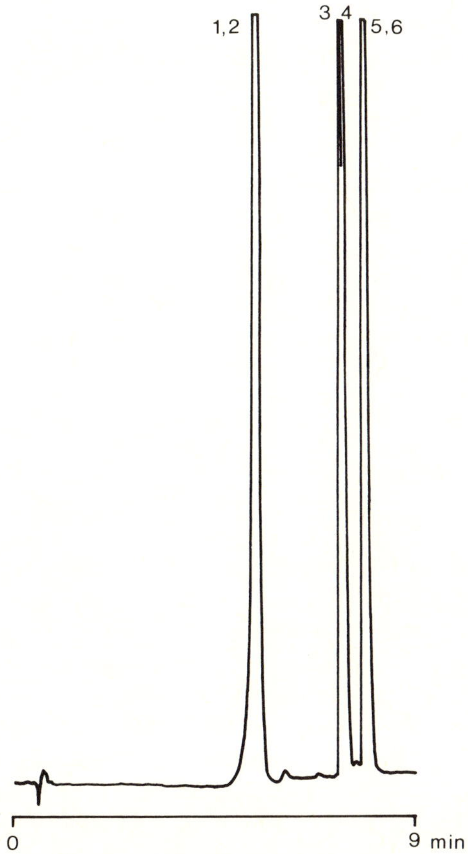

Figure 6.16. Initial gradient separation of the six-component test mixture under control of FASTOPT

Using again the six-component mixture, the performance of FASTOPT can be examined; Table 6.2 summarizes the essential details of the chromatographic system and the objectives that were set for the optimization. Figure 6.16 shows the initial gradient separation of the mixture carried out with a linear gradient from 0 to 100% methanol over 7.5 min. Four peaks are evident, the first at 5.3 min and the last at 8.0 min. For the chromatograph and column being used, these times set a constrained region of 44 to 65% of water to be investigated to achieve the target of all peaks within 1–4 min.

After 23 experiments, the program halted as the simplexes were no longer moving, indicating optimum proportions of water/methanol/acetonitrile of 50:15:35. The movements of the simplexes are shown in Figure 6.17 and the optimized separation in Figure 6.18. Note that the time limit has been exceeded in order to maximize the resolution between all peaks. Higher weightings to the time parameters of the CRF (e.g. 2) would have ensured a closer adherence to the time constraints but at the expense of peak resolution.

The unattended optimization of a reversed-phase separation using identical chromatographic hardware to that used in the example above but with different software has been described.[18] A different CRF was also used:

$$\mathrm{CRF} = \Sigma \ln(f/g) - 100(M-N) \tag{6.7}$$

where N is the number of peaks found in the chromatogram and M is the number of peaks that are expected; f and g refer to valley to baseline distances—the peak separation function of Kaiser described in Chapter 2. This CRF perhaps detracts from one of the great advantags of the simplex procedure, that is that

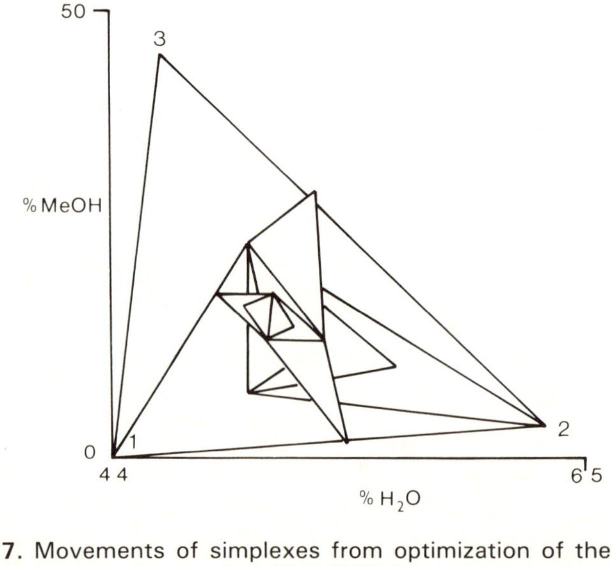

Figure 6.17. Movements of simplexes from optimization of the six-component separation under control of FASTOPT. The initial simplex is 1, 2, 3

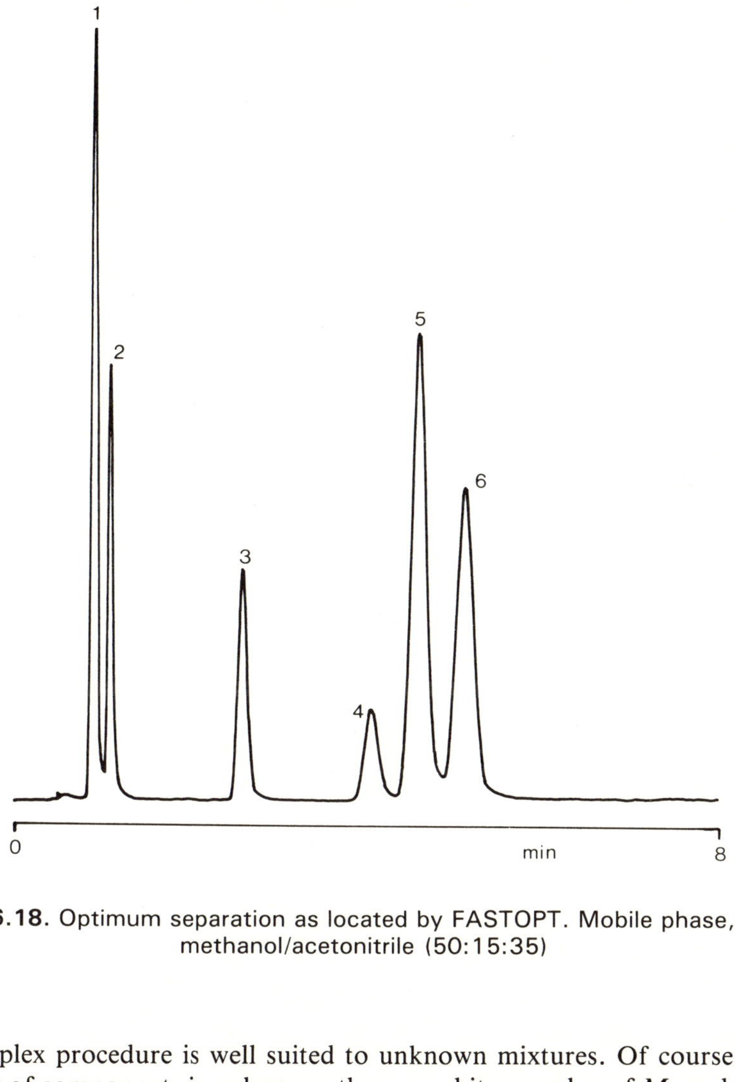

Figure 6.18. Optimum separation as located by FASTOPT. Mobile phase, water/methanol/acetonitrile (50:15:35)

the simplex procedure is well suited to unknown mixtures. Of course, if the number of components is unknown, then an arbitrary value of M can be used.

There are many other examples of the use of the simplex method, the simplex procedure for HPLC optimization being well represented in the literature. Table 6.3 lists a selection of published examples with their references. It is interesting that, for a method originally conceived for automated applications, only the last nine entries refer to on-line computer-controlled optimization.

6.2.5. Software

For such a general technique, used widely in all branches of analytical chemistry, it is perhaps surprising that listings of computer programs for simplex

TABLE 6.3. HPLC optimization using the simplex procedure

Application*	Variables	Ref.
R-P separation of phospholipids	Mobile phase composition	28, 38
Ion-exchange separation of inorganic cations	Binary mobile phase composition	27
R-P separation of PTH-amino acids	4 gradient parameters and flow rate	29
Separation of nucleotides	Buffer concentration, pH, organic modifier level	30
R-P separation of steroids	Temperature, flow rate, binary gradient shape	31
R-P separation of alkaloids	Quaternary mobile phase composition and flow rate	39
R-P separation of dinitrophenyl-hydrazone derivatives	Flow rate, binary gradient parameters	40
R-P separation of nitroaromatics and flavones	Binary mobile phase composition	41
R-P separation of flavones	Binary mobile phase and temperature	42
Normal-phase separation of carotenoids	Ternary mobile phase composition	43
Comparison with mixture designs	Composition, pH, buffer	44
Fully automated examples		
Isocratic R-P separation of pyridines	Binary mobile phase composition and flow rate	32
Gradient R-P separation of antioxidants	3 gradient parameters	32
Isocratic R-P separation of substituted aromatics	Ternary mobile phase composition	32
Isocratic R-P separation of imidazoles and furans	Ternary mobile phase composition	45
Isocratic normal-phase separation of aromatics	Ternary mobile phase composition	32
R-P ion-pair separations of organic bases and organic acids	Ternary mobile phase, (ion-pair reagent, pH), flow rate and temperature	35
R-P gradient separations	Ternary gradient parameters	46
Rapid R-P isocratic separation of sulphonamides	Ternary mobile phase composition	36
Isocratic R-P separation of PTH-amino acids	pH and organic modifier content	18

*R-P = reversed-phase.

optimization are so difficult to find. There are papers that present flow charts describing the particular implementation of the algorithm which they have used[30] and/or methods of carrying out the necessary calculations.[20,47–50] There is a book entirely devoted to programs written in BASIC[51] for carrying out many different types of optimization, which does contain a listing of a program for the Nelder and Mead simplex method. A FORTRAN listing of a program which uses the modified simplex procedure to fit Gaussian shapes to high-performance liquid chromatograms has been published[52] and no doubt this could be adapted to more general applications. Some of the micro- and minicomputer manufacturers can provide simplex optimization algorithms for their computers and the increasing interest in the use of the procedure has prompted one publisher to offer versions of the method aimed at different applications.[53]

The simplex algorithm has also been incorporated into two commercially available liquid chromatographs[54,55] for the unattended optimization of HPLC separations. Whereas one[54] is in compiled BASIC and is not easily accessible to the user, the other[55] is in a relatively straightforward BASIC dialect which can be modified easily by the chromatographer. However, it is apparent that interest in the use of the simplex procedure is increasing but that many will be deterred by the difficulties in obtaining software. A simple implementation of Nelder and Mead's modified simplex procedure is therefore presented as Appendix V. The software listing is actually the kernel of the programs successfully used for automated optimization.[32,33,35,36,45,46,50]

6.2.6. Conclusions

There is little doubt that directed search methods provide a valuable tool for the automated optimization of HPLC separations. They have the great advantages of being intrinsically appealing and readily understood even by those not possessing a great theoretical knowledge of chromatographic principles and techniques. Further, the methods can be adapted to the problems at hand and require little or no knowledge of the sample under investigation. The 'adaptive intelligence' approach is a simple and straightforward entry into these methods but suffers from some limitations on speed and versatility. These limitations are more than compensated for with the sequential simplex procedure, a technique of proved pedigree and applications in all forms of HPLC.

Advantages of directed search methods

1. Will work with completely unknown mixtures.
2. Component identification is not required.
3. Requires only that chromatograms be ordered in relative terms according to their 'goodness'.
4. Can handle many interdependent variables simultaneously.
5. Conceptually simple.
6. Relatively easy to program and implement.
7. Self-correcting if errors occur.
8. New variables can be introduced at any time if they are subsequently considered to be important.

Against

1. No guarantee that the global optimum will be located.
2. Experiments rather than calculates—no predictive ability.
3. May require many experiments (typically 20–40)
4. May explore regions which are of little interest to the chromatographer.
5. Require an appropriate optimization criterion or function.
6. Can be perturbed by experimental noise or a complex or poorly defined response surface.

6.3. MIXED LATTICE AND ITERATIVE METHODS

6.3.1. Introduction

The simplex method just described assumes nothing about the behaviour of the solutes and does not take into account any knowledge of the chromatographic system. The ability to explore a separation without prior knowledge is extremely valuable. It is, however, somewhat inefficient. The methods described in Chapters 3, 4 and 5 are potentially more efficient but their success is largely dependent on there being a suitable model available which describes solute retention as a function of the variables under investigation. If an approximate model could be found which could then be modified during a sequential optimization procedure, this would provide an ideal combination of a model-dependent optimization method with a model-independent method.

6.3.2. Theory

The strategy involved in this mixed experimental design is to include the best aspects of *simultaneous* optimization methods with *sequential* optimization methods. The approach to be described has evolved and been refined substantially over just a few years and it is interesting to see how it has developed. The developments have reflected both chromatographers' needs in dealing with unknown mixtures and also the developments which have occurred in the technology of HPLC. The basic philosophy of the approach is to approximate the response surface with a simple model and then explore the indicated optimum region with an iterative method, refining the model according to the experience gained. The early procedure, for reversed-phase chromatography, has been summarized as follows:[56]

i. The sample is chromatographed using a gradient from 100% water to 100% methanol. From the chromatogram the required solvent strength for isocratic elution can be calculated. This is achieved by using one of the methods described in Chapter 3. A decision can also be made as to whether an isocratic separation will actually be possible within a reasonable time scale.

ii. In accord with the recommendations of the solvent selectivity triangle, the sample is run with isocratic binary mixtures of methanol/water, acetonitrile/

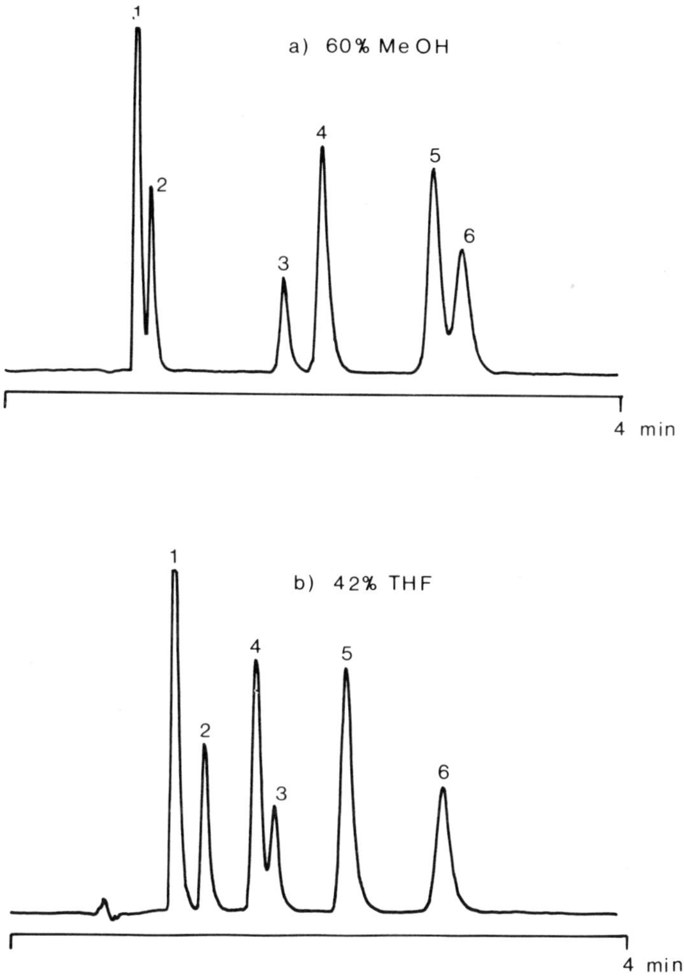

Figure 6.19. Isoelutropic separations of the six-component test mixture

water and/or tetrahydrofuran/water that are predicted to give approximately equal analysis times. The positions of the peaks in the three chromatograms are identified.

iii. Linear plots of the logarithms of the capacity factors versus the composition are constructed for the entire range of ternary mixtures composed of the pairs of limiting binary mixes. These four steps represent the simultaneous part of the experimental design.

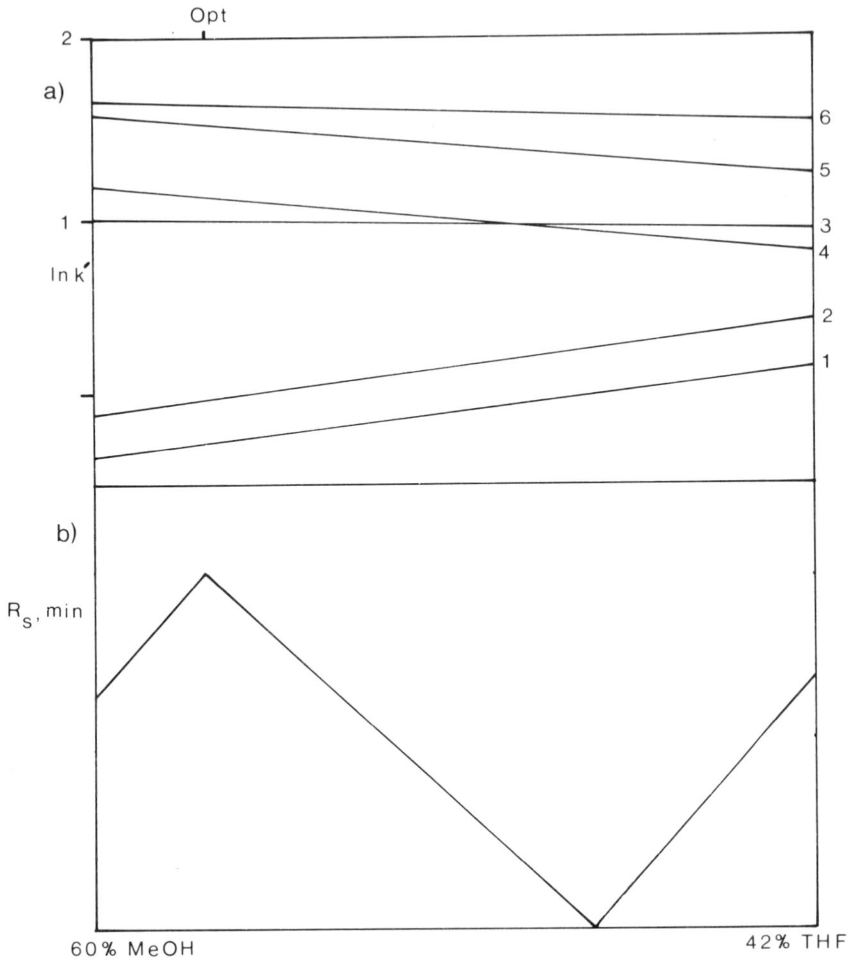

Figure 6.20. Plots of (a) ln k' and (b) minimum resolution against solvent composition for blends of two mobile phases as in Figure 6.19

iv. The sample is then run with the predicted optimum composition to verify the quality of the chromatogram. The calculations and plots are refined, should the separation not be fully optimized, and repeated until the best separation is obtained or until no further improvement can be achieved. This step represents the iterative aspect of the method.

The six-component mixture will again be used to see how this procedure can work in practice. The initial gradient separation indicated that an isocratic mobile phase of methanol/water (60:40) will elute all components in a time of 3 minutes this is the separation shown in Figure 6.19a. Selecting tetrahydrofuran as an alternative modifier, a mobile phase consisting of THF/water (42:58) elutes all

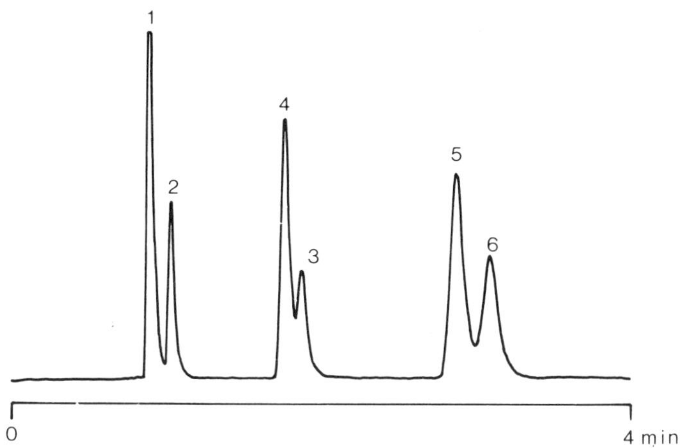

Figure 6.21. Chromatogram obtained at the composition indicated as optimum in Figure 6.20

six components in approximately 3 minutes; this is the separation shown in Figure 6.19b. If linear plots of the logarithms of the capacity factors versus the proportions of these two solvent mixtures are constructed using the data from these two chromatograms, the diagram shown in Figure 6.20a is produced. Also included in Figure 6.20b is the plot of the resolution product. From this plot, a composition of approximately 84% of the methanol solvent and 16% of the THF solvent blend is indicated. The chromatogram shown in Figure 6.21 was obtained using this solvent composition. It is immediately clear that this separation is far from optimum. The data provided by this experiment are now fed back into the procedure and a new plot constructed as shown in Figure 6.22a. From this plot it appears that there is a fairly broad range of compositions which will provide an adequate separation but that the optimum composition contains about 65–70% of the methanol/water solvent blend. Figure 6.23 is the chromatogram obtained at this new indicated optimum composition and, indeed, all peaks are satisfactorily resolved. Just four experiments were required to establish an optimum ternary mobile phase and the optimization can be halted at this point since it is not necessary to explore other solvent compositions—the goal of a complete separation has been achieved. Although in this example the procedure was carried out graphically, it is evident that such a straightforward method can, and is ideally suited to, be implemented on a computer.

For the exploration of all ternary compositions, a minimum of five chromatograms will be required and the whole procedure lends itself well to automated execution, *provided* that solutes can be identified in successive chromatograms. The decision as to the 'best' separation results from the consideration of an optimization criterion. The evolution of the optimization

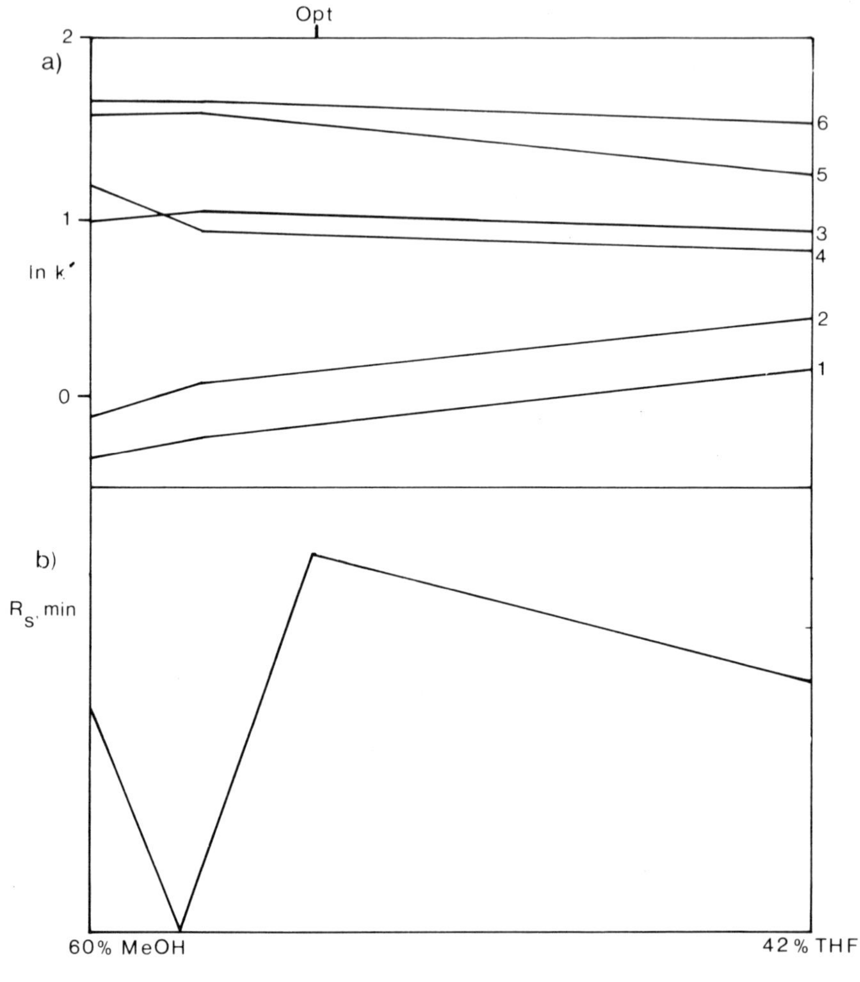

Figure 6.22. Revised plots of (a) ln k' and (b) minimum resolution against composition now incorporating data from Figure 6.21

criterion that is described below has been fully dealt with in Chapter 2 and will not be repeated here: it is the function that has been refined for use with the iterative mixture design method by its originators. The final form is

$$r^* = \prod_{i=1}^{n-1} R_{s_{i+1,i}} \Big/ \left[\left(\sum_{i=0}^{n-1} R_{s_{i+1,i}} \right) / (n-1) \right]^{n-1} \tag{6.8}$$

where R_s is the resolution between the ith peak pair and n is the number of peaks. In the denominator of this equation

$$R_{s_{1,0}} = \sqrt{N}\,(t_1 - t_0)/2(t_1 + t_0)$$

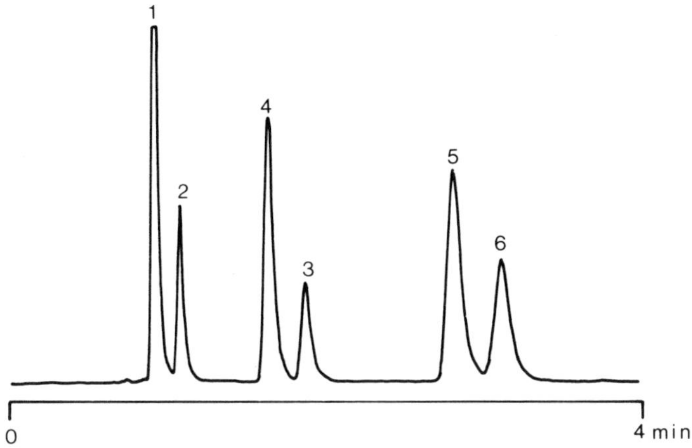

Figure 6.23. Chromatogram obtained using optimum composition from Figure 6.22

where t_1 is the retention time of the first peak and t_0 is the desired minimum retention time, possibly $k'=1$ or $k'=0$. The criterion described by equation 6.8 is designed to provide an even distribution of all peaks early in the chromatogram.

The search for the optimum ternary composition starts from considering the two most promising binary compositions. This range of solvent composition is described by a parameter x having the values $x=0$ at one composition and $x=1$ at the other extreme. It is initially assumed that each solute obeys a linear dependence of $\ln k'$ on composition. The optimization criterion is used to predict a new solvent composition but, rather than run experiments at exactly the composition predicted, allowance is made for the real-life probability that a linear $\ln k'$ dependence is unlikely. This allowance also reduces the chances of experiments being conducted at closely spaced points and therefore improves the efficiency of the searching. For example, if two compositions x_1 and x_2 have previously been investigated, the optimization criterion may predict a new composition x lying between these values. Instead of carrying out an experiment at x exactly, a new, shifted composition x' is investigated, defined by[57]

$$x'=x+0.4[0.5-(x-x_1)/(x_2-x_1)] \tag{6.9}$$

At each new experimental point, the $\ln k'$ plots are refined, the resolution product criterion recalculated and a new optimum composition estimated. If the first selected pair of binary solvent mixes does not provide an acceptable separation, exactly the same method is used to explore the next pair of binary mixes and finally the third pair. In this way it is relatively easy to explore all possible ternary mixtures until such time as the optimum or best separation is realized.

In essence, this iterative approach considers the optimization of ternary mobile phases as a single parameter optimization. It is initially assumed that the retention of solutes in binary iso-elutropic mobile phases, expressed as ln k', can be described by linear segments and this allows prediction of retention by using a straight line. As the procedure continues, this initial straight line is segmented until the real curvature of retentions is adequately described and the procedure can be stopped. It is not necessary to restrict the method to ternary mobile phases formed from mixtures of the two limiting binary solvents (one parameter) as it is possible to use an analogous approach to search a mixture of all three to give quaternary mobile phases.[57] This can be thought of as a two-parameter optimization since it is the mixing ratio of two of the three ternary combinations which can be changed simultaneously.

Ternary solvent mixtures can be represented graphically by a line and quaternary mixtures by a triangle: binary mixtures form the apexes, ternary mixtures the sides and quaternary mixtures are the area inside the triangle. For quaternary optimization the retention behaviour of the solutes is represented initially by a plane fitted through three data points. The philosophy of the quaternary optimization parallels that of the ternary method, i.e.:

i. Each new experiment is used to refine the model to provide a better approximation of the real retention behaviour.

ii. The response surface, as defined by the chosen chromatographic response criterion, is calculated over the whole of the available experimental region to locate the optimum separation.

iii. Having carried out a new experiment, a new composition is calculated at a composition which is 'shifted' from the predicted optimum (see equation 6.9).

iv. The optimization procedure continues until a suitable stop criterion is met.

As each new experiment is conducted, the initial triangle is divided by this data point into three smaller triangles. The entire triangle is thus successively split into a collection of ever smaller triangles. At each stage it is necessary to evaluate the response surface for the current triangle and this consumes large amounts of computer time. For a lattice of 5000 points a computation time of upwards of 30 min is recorded[57] for a fairly powerful minicomputer. The demands on computing time will probably limit this method to consideration of just two parameters, although it is feasible to extend the approach to three parameters. However, it is likely that the law of diminishing returns will apply. There are very few examples of separations for which quaternary mobile phases provide significantly better separations than those which are achievable with less complex mobile phases. The two parameter method does provide, however, a route to the optimization of separations requiring pH or ion-pair manipulation and an example is described in the section on applications.

This procedure as described is relatively fast (particularly for ternary solvent mixtures) and relatively easily adapted to automated optimization. It is not restricted to the optimization criterion described—others could be used to reflect specific problem or instrument needs. It is unfortunate that it does, however, require the identification of solutes if the full benefits of the method are to be realized.

6.3.3. Applications

Applications of the iterative mixture design method have largely been restricted to synthetic mixtures designed to establish the feasibility of the procedure.[57] Thus an early example can be found in the optimization of the separation of six aromatic solutes described by Schoenmakers *et al.*[56] This work was extended and improved[58] to include the optimization criterion described by equation 6.8 for the separation of five diphenylamines. Whereas the earlier method had needed ten chromatograms to define the optimum composition, the new procedure required only seven. Both papers refer to software and programs that were used but a full reference is not included, nor are there any flow charts or listings which might be translated by those so inclined. Both of these applications also refer to the use of absorbance ratioing as an aid to peak identification. This method is not without its drawbacks.[59] Although, in principle, dual-wavelength absorbance ratios should be applicable to solute identification in lattice designs, practical considerations limit their use. The most severe restriction results from the influence of mobile phase composition on the UV spectra but factors such as baseline drift and offset, peak tailing and unresolved peaks also limit this method. The newer diode-array detectors offer a much greater chance for on-line solute identification. Their use in lattice design optimization schemes has been discussed[57,60] and compared with dual- and multi-wavelength absorbance ratioing. The additional advantage that accrues from the use of diode-array detection is that, by using suitable peak deconvolution techniques (Chapter 7), multiple peaks can be resolved into their individual profiles. This can give an estimate of the number of solutes contained in one chromatographic peak, but only if some knowledge of individual components is available of if total overlap has not occurred.

The method has found an interesting application in the optimization of a separation of a synthetic mixture of aromatic solutes carried out on dynamically modified silica.[61] The procedure is entirely analogous to that used for reversed-phase chromatography: all mobile phases contained 0.2 M phosphate buffer (pH 7.5) and cetyltrimethylammonium bromide. Just three experiments provided an acceptable separation of the seven components of the mixture.

In chromatographic optimization, pH has been classified as a discrete variable and has frequently been optimized using factorial designs. Factorial designs suffer from the disadvantage of requiring considerable experimentation. Additionally, it is not really appropriate to optimize pH independently of the other mobile phase components, since solute retention will be dependent on both. The mixed lattice design can be used in the simultaneous optimization of both mobile phase modifier and pH, and this application has been described by Haddad *et al.*[62] Using a mixture of five aromatic acids (whose separation had previously been investigated[63] using a window diagram technique), a

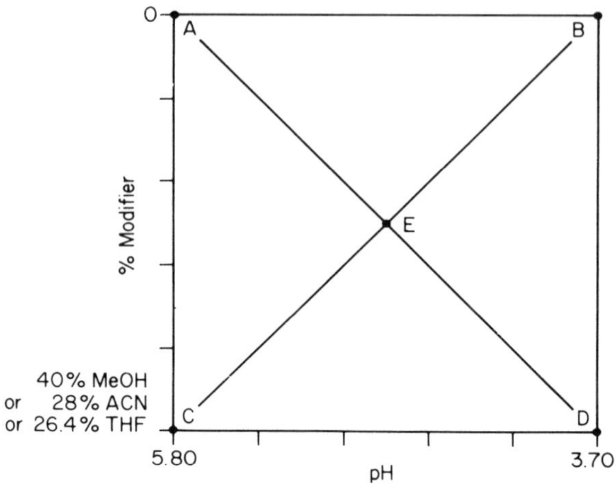

Figure 6.24. Experimental design for combined optimization of pH and organic modifier content of the mobile phase in reversed-phase chromatography. The letters A–E represent the mobile phase compositions used to provide the experimental data necessary to initiate the optimization procedure. *Reproduced from ref. 62 with permission of Elsevier Science Publishers BV*

mixture design approach was applied first to the optimization of the pH as a single parameter. This procedure is entirely analogous to that already described for modifier content optimization.

This design was extended, as outlined above, to allow two-parameter optimization and was used for the simultaneous optimization of pH and organic modifier. The experimental design in this case requires five initial chromatograms for each modifier and, in the example given, they were carried out as shown in Figure 6.24. Instead of plotting ln k' versus modifier content, a planar surface is constructed over the desired range of pH and modifier content. Each plane interconnects three experimentally derived retention values and the planes are used in the calculation of the values of the optimization criterion. The mid-point of the experimental design, point E, is included so that the total optimization area can be subdivided into four retention planes. These four planes are modified subsequently by the inclusion of more experimental data acquired under the iterative part of the procedure. Figure 6.25 shows the three-dimensional representations of the relative resolution product (equation 6.8) over the ranges of pH and modifier content described by Figure 6.24, together with chromatograms obtained under predicted optimum conditions. For each individual modifier, the optimum composition was finally located after three additional experiments, requiring a total of eight experiments. This amount of experimentation compares well with the 16 experiments that would have been required for a full, two-level, factorial design. Inspection of the chromatograms reveals that the optimized separations reflect the aim of the optimization criterion, namely to disperse the peaks as evenly as possible throughout the

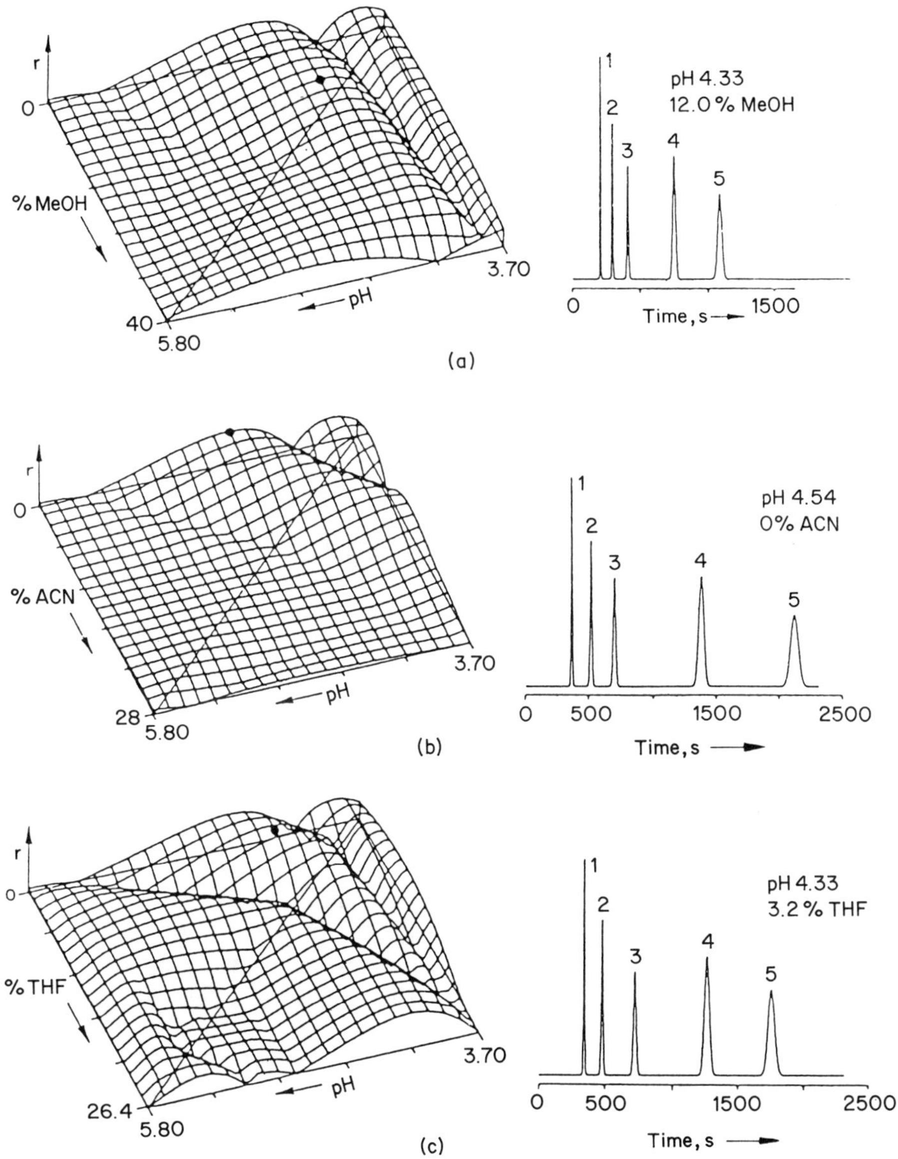

Figure 6.25. Combined optimization of mobile phase pH and organic modifier by means of the relative resolution product criterion (*r*). Each three-dimensional surface illustrates the variation of *r* over the indicated mobile phase compositions. Chromatograms obtained with the selected optimum mobile phase compositions (indicated by the black spots) are also shown. The organic modifiers used were as follows: (a) methanol, 0–40%; (b) acetonitrile, 0–28%; and (c) tetrahydrofuran, 0–26.4%. Column, 150 cm × 0.46 cm i.d. containing 5 μm Hypersil ODS. Solutes: 1 = 1,4-benzenedicarboxylic acid; 2 = 4-aminobenzoic acid; 3 = 4-hydroxybenzoic acid; 4 = 2-aminobenzoic acid; 5 = benzoic acid. *Reproduced from ref. 62 by permission of Elsevier Science Publishers BV*

chromatogram without particular reference to the separation time. However, in the case of the methanol system, most workers would wish to speed up the separation. This need was recognized by the authors and the separations further investigated by adding a 1/time term and specifying a desired minimum resolution product of 2. By considering the time requirement after first locating regions of optimum composition, it is argued that problems sometimes encountered by the time consideration over-influencing the result of the optimization are eliminated. The results obtained in this sequential search step are illustrated in Figure 6.26. The surface is flat ($1/t$ is set to zero) for all regions where the resolution criterion is not satisfied but reflects the value of $1/t$ in the other areas. The highest point is the fastest separation consistent with the resolution constraint. The chromatograms now achieved at the optimum conditions are considerably faster than those illustrated earlier, but do not compromise resolution unduly.

The extension of the lattice design to reversed-phase ion-pair separations has also been described.[64,65] Catecholamines and metabolites were investigated and the method used was equivalent to that described, using an iterative procedure with interconnecting linear segments. To identify the individual solutes, a diode-array detector was used. It was necessary to add principal components analysis to cope with the identification of co-eluting solutes and this obviously adds to the complexity of the whole procedure, requiring some sophisticated software and raising some doubts as to the success rate for totally unattended optimization.

A computer-assisted mixture design method was used by Haddad and Cowie[66] to select the optimum eluent composition in single-column ion chromatography. The first step was to optimize the phthalate buffer concentration at a constant pH, and success here relied on the linear relationship between log k' and log [eluent]. The second step was to consider the optimization of the eluent pH at constant buffer concentration. In this situation a linear relationship does not exist between log k' and the eluent pH. The iterative aspect of the mixture design approach means that it can be applied successfully in these cases since each experiment provides more data that can be progressively used to more closely approximate the true retention relationship. It is likely that the method described in ref. 61 will provide the basis for the simultaneous optimization of both pH and buffer concentration and prove to be a useful tool in the optimization of ion chromatographic separations.

As with so many optimization schemes, success rates also seem to be dependent on the complexity of the mixture. The lattice design approach was found[67] to work adequately for a mixture of 21 monohydroxy aromatic acids, although all peaks were not fully resolved. (The investigators also tried window diagrams with the result that an alternative optimum was suggested, but this was actually inferior.)[68]

6.3.4. Summary

The mixed lattice and iterative methods form a useful hybridization of formal experimental designs with methods providing real-time feedback to the

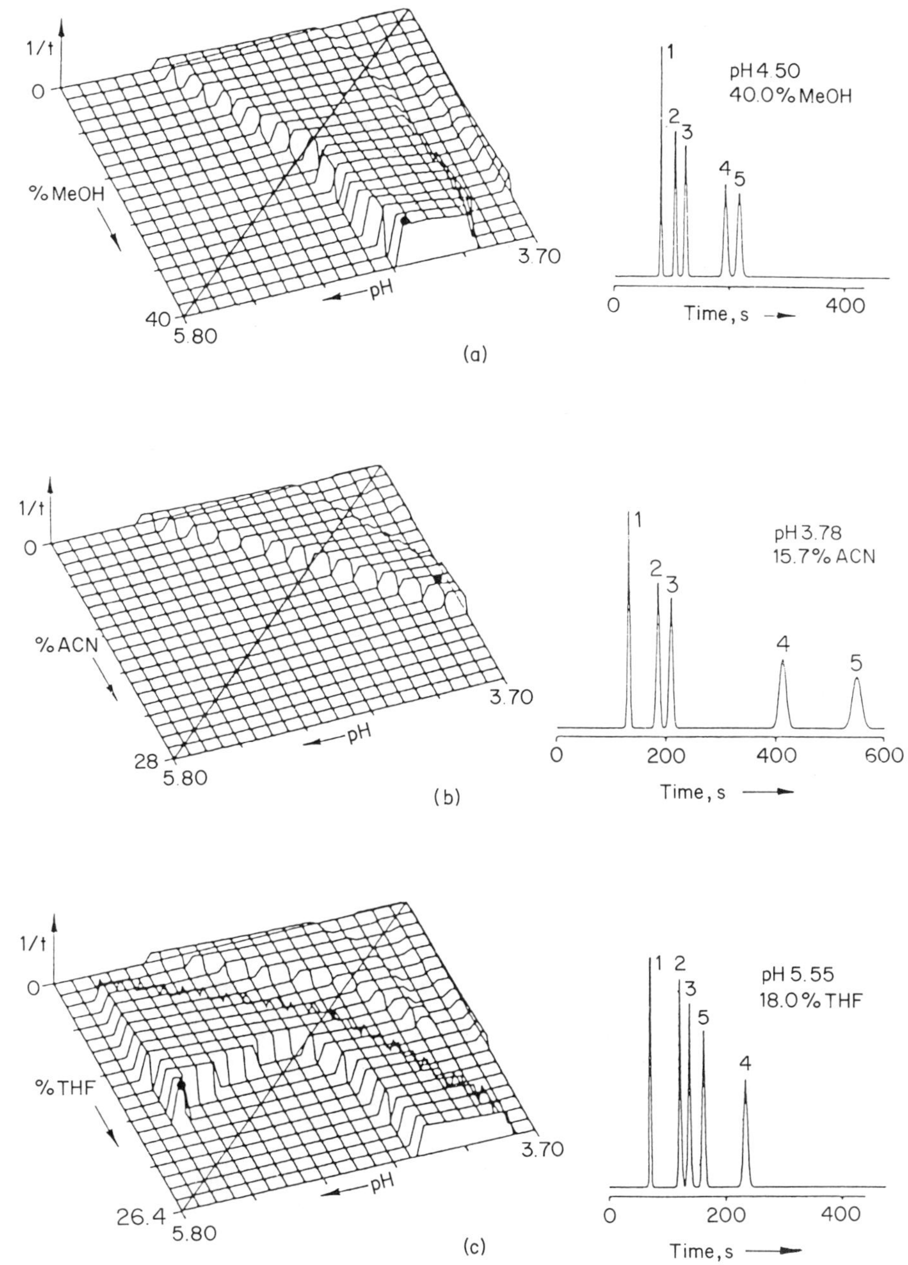

Figure 6.26. As for Figure 6.25, except that the sequential resolution–time procedure was used and the surface illustrates the variation of $1/t$ (where t is the retention time of the most retained peak). A value of 2 was used for the minimum resolution requirement. *Reproduced from ref. 62 by permission of Elsevier Science Publishers BV*

optimization procedure. Solutes are assumed to be well behaved, but ideal behaviour is not a prerequisite. The successive approach of optimizing first binary, ternary and ultimately quaternary solvent mixes makes it possible to continue the procedure only so long as to provide an acceptable separation. In most cases it should be possible to locate an optimum separation within 5–10 experiments.

For iterative mixture designs

1. Generally efficient, with 5–10 experiments.
2. Ideal solute behaviour not assumed.
3. Versatile in mobile phase parameters (e.g. modifiers, pH. ion-pair reagent).
4. Versatile in optimization criteria—the user can decide on an appropriate criterion.
5. Suitable for up to two parameters simultaneously (i.e. up to quaternary solvent mixtures).

Against

1. Solutes need to be identified in successive separations.
2. Demanding on computing time for more than one parameter.
3. Complex software for fully automated use.

6.4. REFERENCES

1. M. P. T. Bradley and D. Gillen, *Spectra-Physics Chromatogr. Rev.*, **10** (2) (1983) 2.
2. M. P. T. Bradley and D. Gillen, *Spectra-Physics Chromatogr. Rev.*, **11** (1) (1984) 10.
3. J. W. Dolan, J. R. Gant and L. R. Snyder, *J. Chromatogr.*, **165** (1979) 31.
4. D. G. Gillen, B. Tomlinson and R. Honganen, personal communication.
5. G. E. P. Box and K. B. Wilson, *J. R. Stat. Soc., B.* **13** (1951) 1.
6. G. E. P. Box, *Appl. Stat.*, **6** (1957) 81.
7. W. Spendley, G. R. Hext and F. R. Hinsworth, *Technometrics*, **4** (1962) 441.
8. B. R. Kowalski, *Trends Anal. Chem.*, **1** (1981) 71.
9. D. L. Massart and G. Hoogewijs, *Pure Appl. Chem.*, **55** (1983) 1861.
10. J. A. Nelder and R. Mead, *Comput. J.*, **7** (1965) 308.
11. P. G. King, *PhD Dissertation*, Emory University, Atlanta, 1974.
12. A. E. Brookes, T. J. Leary and D. W. Golightly, *Anal. Chem.*, **53** (1981) 720.
13. R. R. Ernst, *Rev. Sci. Instrum.*, **39** (1968) 998.
14. M. W. Routh, P. A. Swartz and M. B. Denton, *Anal. Chem.*, **49** (1977) 1422.
15. P. B. Ryan, R. L. Barr and H. D. Todd, *Anal. Chem.*, **52** (1980) 1460.
16. P. F. A. Van der Wiel, R. Maassen and G. Kateman, *Anal. Chim. Acta*, **153** (1983) 83.
17. E. R. Aberg and A. G. T. Gustavsson, *Anal. Chim. Acta,* **144** (1982) 39.
18. J. H. Nickel and S. N. Deming, *Liq. Chromatogr. Mag.*, **1** (1983) 414.
19. L. A. Yarbro and S. N. Deming, *Anal. Chim. Acta*, **73** (1974) 391.
20. D. E. Long, *Anal. Chim. Acta*, **46** (1969) 193.
21. S. N. Deming and S. L. Morgan, *Anal. Chem.*, **45** (1973) 278A.
22. S. L. Morgan and S. N. Deming, *Anal. Chem.*, **46** (1974) 1170.

23. S. N. Deming and L. R. Parker, *CRC Crit. Rev. Anal. Chem.*, **7** (1978) 187.
24. S. N. Deming and S. L. Morgan, *Anal. Chim. Acta*, **150** (1983) 183.
25. P. J. Golden and S. N. Deming, *Lab. Microcomput.*, **2** (1984) 44.
26. S. L. Morgan and S. N. Deming, *J. Chromatogr.*, **112** (1975) 267.
27. R. Smits, C. Vanroelen and D. L. Massart, *Fresenius Z. Anal. Chem.*, **273** (1975) 1.
28. M. L. Raney and W. C. Purdy, *Anal. Chim. Acta*, **93** (1977) 211.
29. M. W. Watson and P. W. Carr, *Anal. Chem.*, **51** (1979) 1835.
30. V. Svoboda, *J. Chromatogr.*, **201** (1980) 241.
31. D. M. Fast, P. H. Culbreth and E. J. Sampson, *Clin. Chem.*, **28** (1982) 444.
32. J. C. Berridge, *J. Chromatogr.*, **244** (1982) 1.
33. J. C. Berridge, *Chromatographia*, **16** (1982) 173.
34. A. F. Fell, *Proc. Anal. Div. Chem. Soc.*, **17** (1980) 512.
35. J. C. Berridge, *Analyst (London)*, **109** (1984) 291.
36. J. C. Berridge and E. G. Morrissey, *J. Chromatogr.*, **316** (1984) 69.
37. J. W. Weyland, C. H. P. Bruins, H. J. G. Debets, B. L. Bajema and D. A. Doornbos, *Anal. Chim. Acta*, **153** (1983) 93.
38. M. L. Raney, S. W. McLear and W. C. Purdy, *Abstracts of 'Biochemische Analytik '74 combined with 1st European Congress of Clinical Chemistry'*, Munich, April 1974.
39. D. L. Dunn and R. E. Thompson, *J. Chromatogr.*, **264** (1983) 264.
40. L. G. Sabate, A. M. Diaz, X. M. Tomas and M. M. Gassiot, *J. Chromatogr. Sci.*, **21** (1983) 439.
41. J. Rafel and J. Lema, *Afinidad*, **41** (1984) 30; *Chem. Abstr.*, **100** (1984) 216141c.
42. J. Rafel, *J. Chromatogr.*, **282** (1983) 287.
43. A. S. Kester and R. E. Thompson, *J. Chromatogr.*, **310** (1984) 372.
44. F. V. Warren and B. A. Bidlingmeyer, Paper presented at 188th ACS Meeting, Philadelphia, 1984, Paper 108.
45. J. C. Berridge, *Proc. Anal. Div. Chem. Soc.*, **19** (1982) 472.
46. J. C. Berridge, Paper presented at Analyticon 83, London, 1983, Paper 505.
47. C. L. Shavers, M. L. Parsons and S. N. Deming, *J. Chem. Educ.*, **56** (1979) 307.
48. D. J. Legget, *J. Chem. Educ.*, **60** (1983) 707.
49. R. Cela and J. A. Perez-Bustamante, *Comput. Appl. Lab.*, **2** (1983) 137.
50. J. C. Berridge, *Microprocessors Microsystems*, **7** (1983) 19.
51. B. D. Bunday, *BASIC Optimisation Methods*, Edward Arnold, London, 1984.
52. S. T. Balke, *Quantitative Liquid Chromatography, a Survey of Chemometric Procedures*, Elsevier, Amsterdam, 1984.
53. INSTRUMENTUNEUP and CHEOPS (CHemometric OPtimization using Simplex), two versions of simplex optimization methods, Elsevier, Amsterdam, 1984.
54. SUMMIT Chromatographic System, Bruker Spectrospin, Coventry, UK.
55. TAMED Chromatographic System, LDC Milton Roy, Stone, UK.
56. P. J. Schoenmakers, A. C. J. H. Drouen, H. A. H. Billiet and L. de Galan, *Chromatographia*, **15** (1982) 688.
57. A. C. J. H. Drouen, 'Computerized Optimization and Solute Recognition in Liquid Chromatographic Separations', *PhD. Thesis*, Technische Hogeschool Delft, 1985.
58. A. C. J. H. Drouen, H. A. H. Billiet, P. J. Schoenmakers and L. de Galan, *Chromatographia*, **16** (1982) 48.
59. A. C. J. H. Drouen, H. A. H. Billiet and L. de Galan, *Anal. Chem.*, **56** (1984) 971.
60. A. C. J. H. Drouen, H. A. H. Billiet and L. de Galan, *Anal. Chem.*, **57** (1985) 962.
61. S. H. Hansen and P. Helboe, *J. Chromatogr.*, **285** (1984) 53.
62. P. R. Haddad, A. C. J. H. Drouen, H. A. H. Billiet and L. de Galan, *J. Chromatogr.*, **282** (1983) 71.
63. S. N. Deming and M. L. H. Turoff, *Anal. Chem.*, **50** (1978) 546.

64. L. de Galan, H. A. H. Billiet and A. C. J. H. Drouen, Paper presented at 8th International Symposium on Column Liquid Chromatography, New York, 1984, Paper 3a–B3.
65. H. A. H. Billiet, A. C. J. H. Drouen and L. de Galan, *J. Chromatogr.*, **316** (1984) 231.
66. P. R. Haddad and C. E. Cowie, *J. Chromatogr.*, **303** (1984) 321.
67. H. A. Cooper and R. J. Hurtubise, *J. Chromatogr.*, **324** (1985) 1.
68. H. A. Cooper and R. J. Hurtubise, *J. Chromatogr.*, **328** (1985) 81.

7 Post-chromatographic Techniques and Future Developments

To conclude this discussion on techniques suitable for the automated optimization of HPLC separations, the question of which of the methods described should be selected will be discussed briefly. Some methods which can reveal further information about the quality of the separation will then be considered. When dealing with an unknown mixture, it is imperative to take advantage of all the potential information. This information may be accessible as a consequence of the inherent detection capabilities that are available with the chromatographic hardware. Linear diode-array detectors obviously fall into this category. Chemometric methods can also be used to reveal additional information about the separation, information which may be impossible to extract without such assistance. Techniques such as differentiation and peak deconvolution fall into this category.

Finally, a tentative—and speculative—peek into the future will be taken. No doubt there will be continuing incremental advances in chromatographic hardware but it will surely be in computational techniques that the major developments will be seen. As developments in chemometrics and in artificial intelligence techniques bring them both ever closer together, there is little doubt that it is only a matter of time before some dramatic capabilities arise out of 'Expert Systems'. Already there are systems available which go a considerable way to directing the chromatographer in the quest for an optimized separation without the need for experimentation.

7.1. SUMMARY AND COMPARISON OF METHODS

When approaching the optimization of an HPLC separation, the basic strategy can be summarized in three steps.

1. Formulate a criterion which expresses the quality of the separation in some manner; ideally, this should be a single number but the criterion will almost

certainly express the evaluation of several responses such as resolution and analysis time.

2. Define the *parameter space* by selecting the number and range of variables; this will be simplified if the parameters are continuous.

3. Choose a procedure to locate the highest (or lowest, depending on the particular criterion) criterion value in the parameter space; this should represent the *global optimum* of the response surface.

The procedures that are available for Step 3, the location of the optimum conditions, can be classified into four categories, according to the number of experiments which might be required.

i. Mapping: this can be univariate, from factorial designs, or simply a 'brute-force' searching of the whole of the available parameter space.

- Uses a pre-determined measurement scheme (therefore inflexible and all experiments may need to be conducted before inferences can be drawn).
- Many data points will be required to gain an accurate description of the response surface (50–150 for a complex surface and three or four variables).
- Not essential to know the elution order.
- Software for experimentation relatively simple.

ii. Self searching: the best example of this category is the sequential simplex design.

- Efficient for the simultaneous optimization of many inter-dependent parameters (usually 15–30 experiments for three or four variables; adding more variables will not usually require a substantial increase in the number of experiments).
- The global optimum may be missed and a local optimum located instead.
- More intricate software is required but it can be easily handled by a microcomputer.

iii. Regression methods: examples are window diagrams, the estimation of the response surface from data acquired following a simplex lattice design (but note the analogy with procedure type i above) i.e. the Sentinel approach.

- Fits a mathematical expression (usually a quadratic) to the response surface.
- Works indirectly from *retention* surfaces.
- Need to know the elution order.
- Number of variables restricted (three is a reasonable maximum).
- Where the elution order is easy to determine, the number of experiments required can be few, typically 5–10; where the determination of the elution order requires standards to be chromatographed, the number of experiments required rises dramatically to 30–60.
- Software complex for full implementation.

iv. Iterative regression: this is the continuous refinement of a linear model until the response surface is adequately described.

- Approaches the true retention/response surface stepwise.
- Ideally requires the elution order.
- Restricted to three independent variables (i.e. a quaternary mobile phase).

• As in iii, where the elution order is easy to determine, as few as 3–5 experiments may suffice.

• Software for three independent variables is very complex and is likely to be beyond the capabilities of many microcomputers.

These categories have been arranged in order of decreasing number of experiments required to locate the optimum separation. Unfortunately, to reduce the number of experiments, it is usually necessary to have more information to hand. For example, iterative regression methods are most successful when all peaks are known and can be identified in successive chromatograms. As new techniques with multi-channel detectors are developed, this requirement may be reduced as peaks of unknown origin can be tracked on the basis of their spectroscopic 'fingerprints'. It is also unfortunate that, going through the categories, the complexities of the software requirements also increase. Thus it is comparatively easy to write software to carry out a 'brute-force' search of a parameter space but the software to undertake the iterative regression optimization of three independent variables (complete with peak tracking) will be far from easy! Nevertheless, software for all categories is becoming more widely available. The recommendation must be to use the scheme of the highest category consistent with the information known about the mixture and the software available for the chromatograph.

7.2. POST-CHROMATOGRAPHIC METHODS

7.2.1. Diode-array Detection

The enormous data-capture capabilities of photodiode-array detectors present the chromatographer with a wealth of information for subsequent analysis.[1–8] As well as the traditional two-dimensional presentation of detector signal against time (Figure 7.1), it is possible to display the UV spectra of the peaks. Such spectra can be captured in 'real time' without the need to interrupt the eluent flow. In Figure 7.1 the spectra obtained on the upslope, the apex and the downslope of each peak are shown. It is apparent that the fourth peak is a composite of at least two peaks as the spectra on the upslope and downslope are completely different and further optimization of the separation is required.

Most diode-array detectors allow the collection of full spectral data at rates of up to several scans per second. With these data it is possible to construct pseudo-three-dimensional or isometric plots of absorbance, wavelength and time. Moreover, these plots can be manipulated to allow the data to be viewed from different angles, including from the end of the chromatogram towards the beginning. Two such plots, depicting an incomplete separation of the six-component test mixture, are shown in Figure 7.2. Pretty though these plots may be, it is difficult to extract quantitative data from them and, to aid in the plotting, the information has had to be condensed to a manageable amount. Nevertheless, such plots have been advocated as rapid and successful means of developing separation strategies in toxicological and metabolic studies.[9–13] A potentially

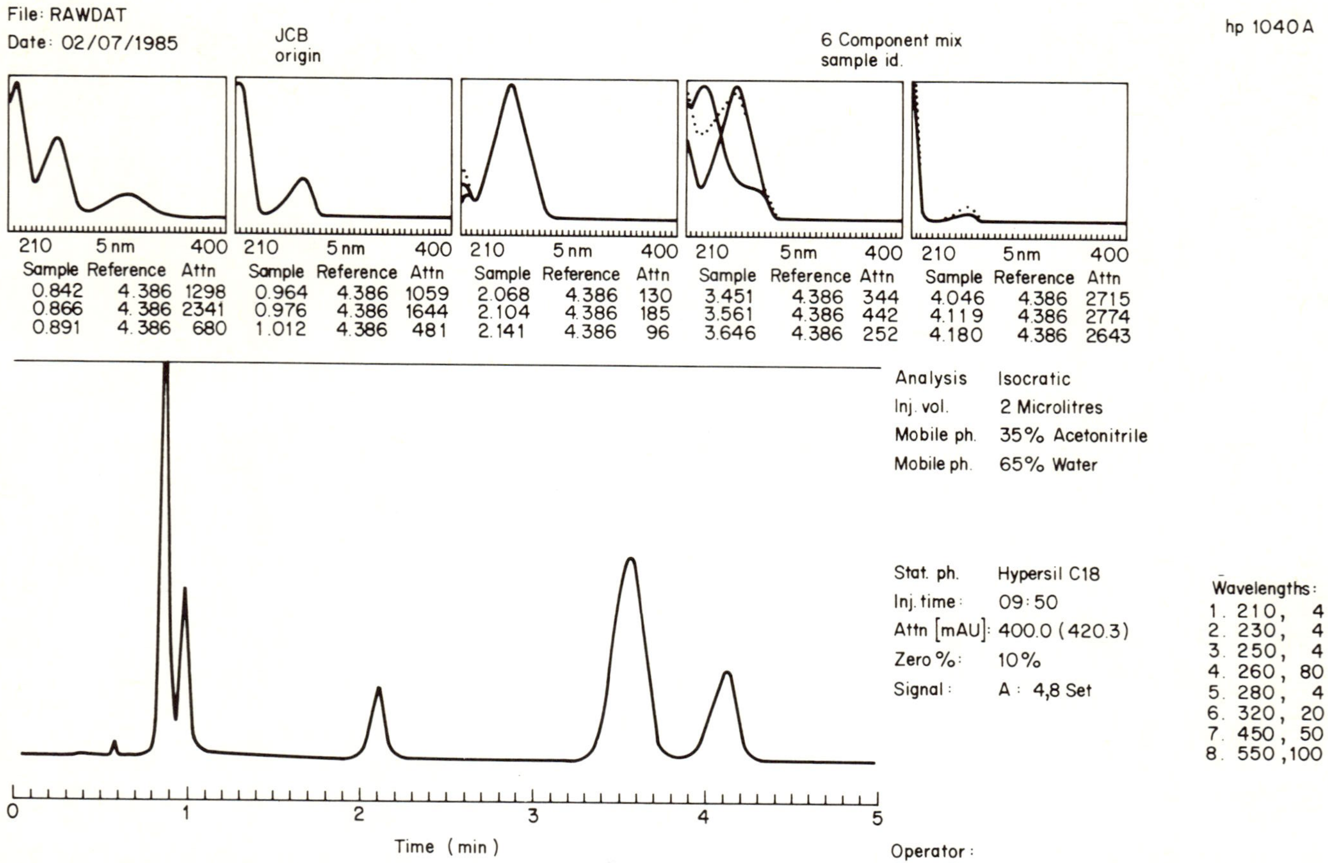

Figure 7.1. Incomplete separation of the six-component mixture showing UV spectra obtained from upslope, apex and downslope of peaks. Note the inhomogeneity of the peak at 3.5 min

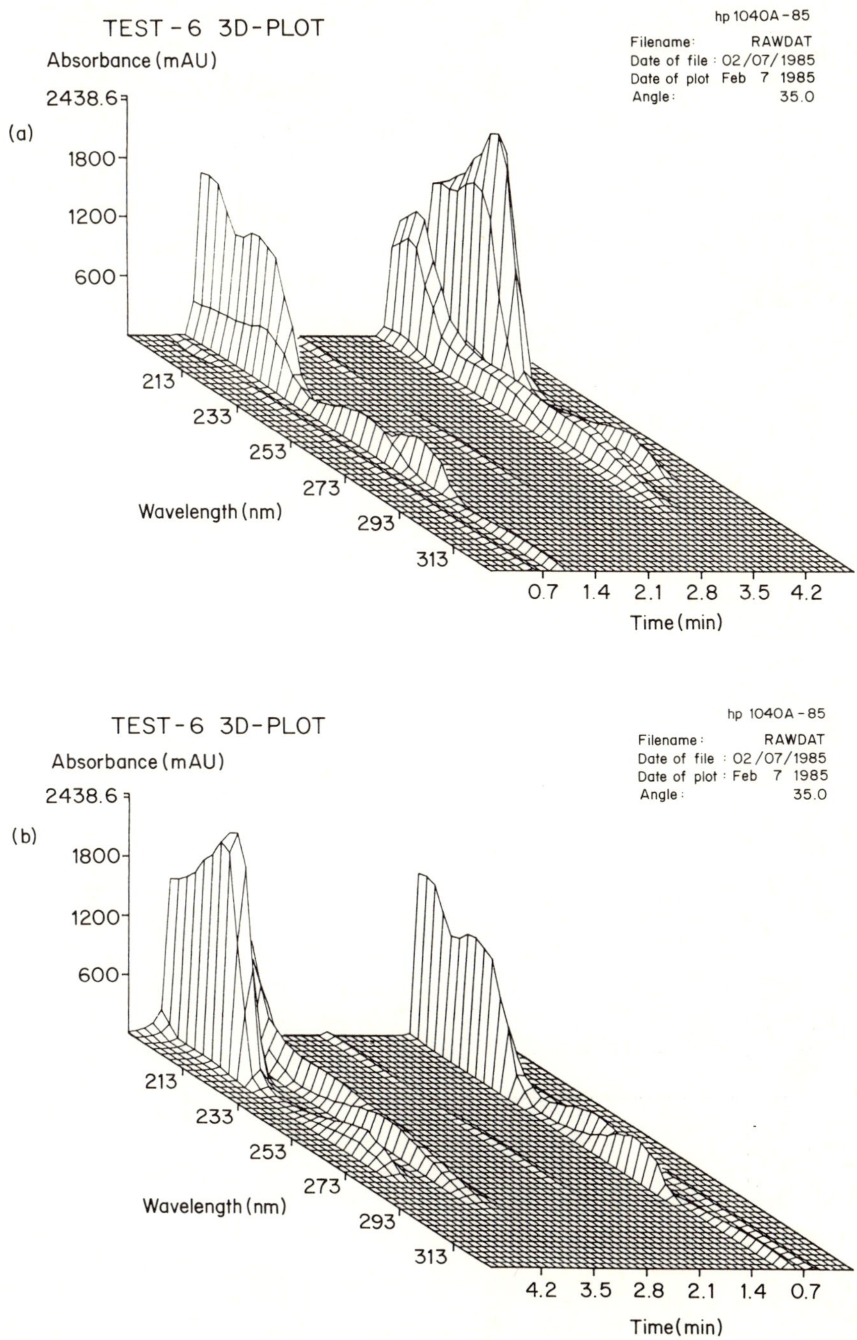

Figure 7.2. (a) Forward and (b) reverse isometric plots of the same separation as in Figure 7.1

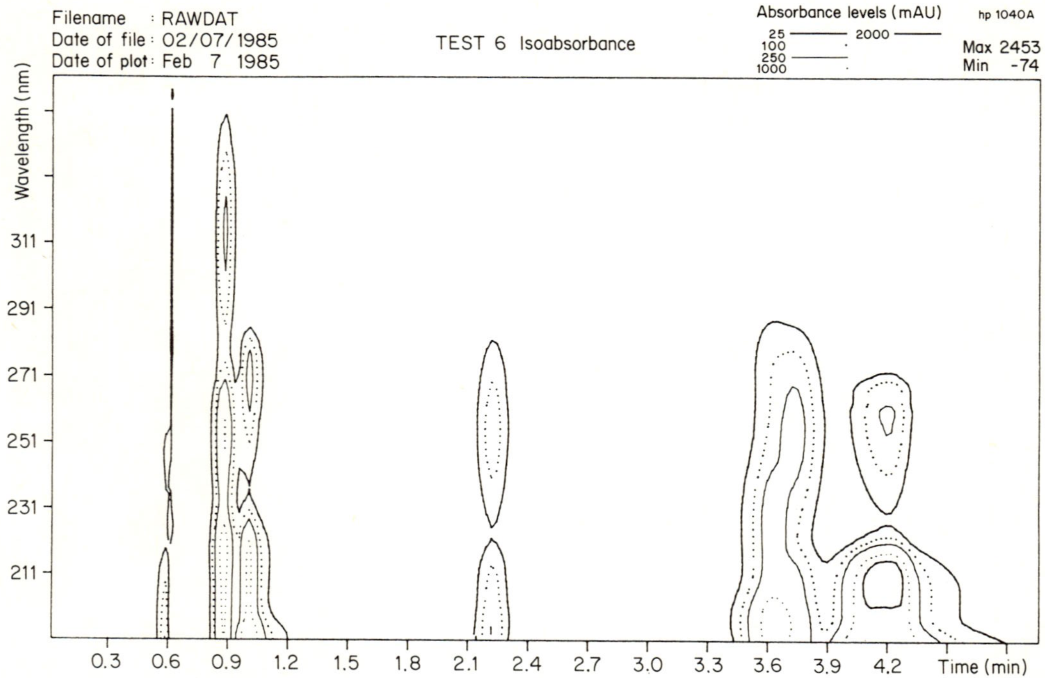

Figure 7.3. Isoabsorbance (contour) plot of same separation as in Figure 7.1

more informative way of presenting the chromatograms is to use the cartographic technique of a contour plot,[14] a map of signal intensity in a wavelength–time domain. Such a map has the property that all components in the 'spectro-chromatogram' that exceed a minimum intensity can be seen simultaneously. An example is shown in Figure 7.3, using the same separation as shown in Figure 7.1. From this plot it is much easier to see the incomplete resolution of peak pairs, in particular the overlap of components at approximately 3.6 min. It is also easy to determine the optimum detection wavelength(s) for a single channel detector, in this case either 205 or 260 nm. The ability to produce such plots rests entirely on the capabilities of the software available for the computer controlling the detector. Some systems may take hours to produce such a plot,[15] whereas others can display them in real time[16] and in eight colours!

Having acquired such a mass of data, there are many digital techniques that can be applied to examine the purity of peak profiles.[1,6,13,17,18] One of the simplest is to plot the ratio of absorbances at two carefully chosen wavelengths as a function of time. Any inhomogeneity is indicated as a discontinuity of the plot, provided that the spectral properties of the overlapping solutes are sufficiently different.[7] Figure 7.4 clearly shows the discontinuity at 3.6 min, confirming the overlap of at least two components at this point. Such ratiograms will not indicate when a total overlap of two peaks has occurred and there are also difficulties to be overcome should there be baseline drift.[19] Comparison of spectra captured on the leading edge, the apex and the trailing edge of a peak can also give an indication of the purity of a peak,[2,20] as is so graphically shown in Figure 7.1

One other method of assessing peak purity deserves special mention. The method known as spectral suppression[18] uses difference spectroscopy to suppress, or subtract, selectively the component of a chromatographic peak which arises from a known constituent. What remains, if anything, will be due to co-eluting peaks. The method relies on the fundamental spectroscopic principle that the absorptivity of a pure component at one wavelength (ϵ_1) is a fixed proportion of its absorptivity at another (ϵ_2):

$$\epsilon_1 = K\epsilon_2 \qquad (7.1)$$

Therefore, provided that the linear range of the detection system is not exceeded, this can be extended to the absorbance relationship:

$$A_1 = KA_2 \qquad (7.2)$$

The absorbance of that pure component can be suppressed simply by computing the function shown as equation 7.3 with respect to time:

$$\Delta A_{1,2} = A_1 - KA_2 = 0 \qquad (7.3)$$

If two wavelengths can be chosen where the absorptivities of the pure component are equal, K will be equal to unity. If this condition is fulfilled, spectral

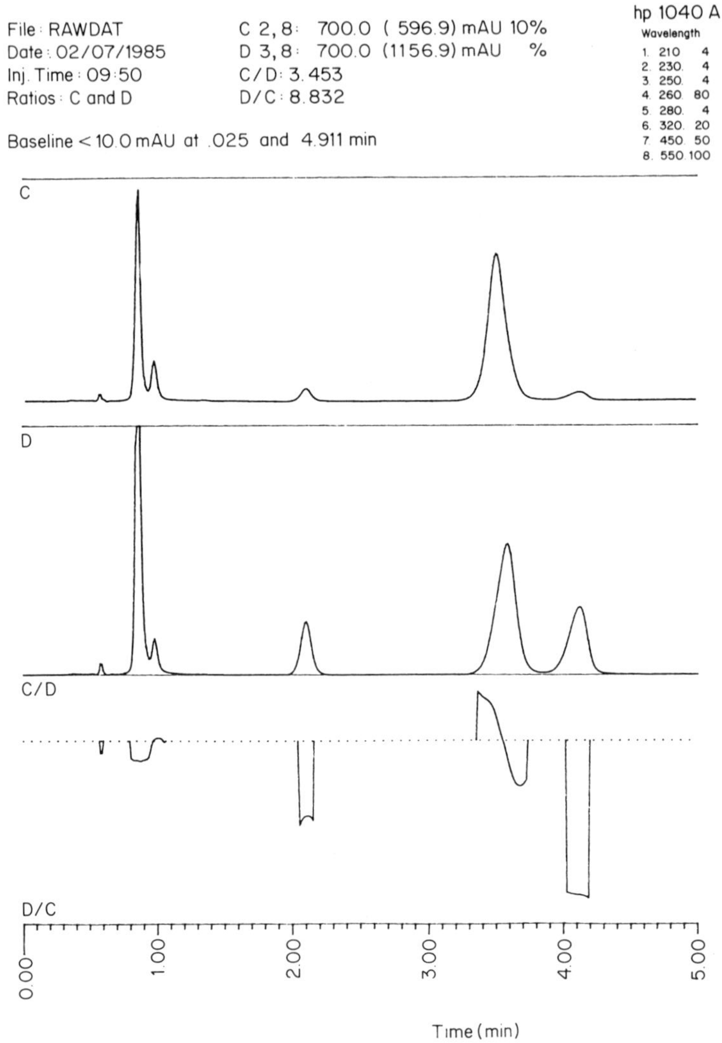

Figure 7.4. Ratiogram of the separation shown in Figure 7.1. C = chromatogram at 230 nm; D = chromatogram at 250 nm

suppression can be carried out relatively easily on most commercial diode-array spectrophotometers without the need to write additional software. However, the value of spectral suppression techniques is rapidly gaining acceptance and suitable software is being increasingly provided by the equipment manufacturers.[16]

The application of spectral deconvolution techniques to examine elution profiles is useful for qualitative and quantitative determinations of known components which are not resolved. It does require that some information on the number of solutes is known beforehand and is therefore not likely to be of great value in determining whether an optimum separation has been achieved. The method may find uses in situations where an unexpected component appears on the leading or trailing edge of what should have been a pure peak[7] and can be used to estimate the spectrum of the unknown component.[21]

The general techniques of *factor analysis* provide methods for extracting information from complex chromatograms. Factor analysis seeks to establish the number and identities of the factors (i.e. variables or parameters) involved and to generate an equation to fit and predict their influences as a sum of linear terms. A detailed description on how to carry out the methods of factor analysis is outside the scope of this book but there are two books to which the reader is specifically referred.[22,23] Factor analysis is an exciting technique to combine with the new multi-channel detectors since it promises to provide a very large amount of information. Computer implementation is essential and, although the software is complex, programs are becoming increasingly available.[23] *Principal components analysis* is likely to be the most favoured method of dealing with co-eluting peaks of unknown complexity.[22–26]

The success of principal components analysis is dependent on the spectral differences between the solutes and the degree of resolution achieved. At present its accurate use is limited to systems consisting of just two or three peaks. Osten and Kowalski[27] have shown that it is possible to resolve co-eluting peaks without any prior assumptions of the underlying peak shape, provided that sufficient dissimilarities exist. The use of diode-array detection and principal components analysis is starting to find application in automated separation optimization[28] as a method to identify and hence track the positions of the individual solutes.

7.2.2. Approaches to the Problem of Assessing Peak Purity Using Simple Detectors

Even without the advantages that accrue from the use of multi-channel detectors, there are techniques which can be applied to peak deconvolution.[29–31] The references of Rosenbaum *et al.*[29] and Balke[31] may be of particular interest for automated applications since they both contain software for curve fitting which could be used on-line. Most methods require that some resolution be discernible, the number of peaks be known and that the peaks can be fitted to Gaussian or modified Gaussian functions. The spectral suppression technique of Carter *et al.*[18] can be used with a single-channel UV detector to provide further information on the purity of chromatographic peaks. If the chromatographic conditions can be maintained sufficiently stable, it is possible to run two chromatograms under identical conditions except for the detection wavelength. With a suitable chromatographic data system a multiple of one chromatogram can be subtracted from the other to eliminate effectively the peak due to a specific component. The use of this method obviously requires a knowledge of the spectral properties of the peak to be eliminated.

Absorbance ratioing is, as has been seen above, a method which can be used for the assessment of peak purity.[32,33] Its sensitivity depends on the magnitude of the differences in absorptivities of the solute molecules at the wavelengths selected and its reliability is heavily dependent on the absence of baseline artefacts.[19] There are non-diode-array detectors commercially available which can monitor at selected wavelengths and thereby provide absorbance ratios. Alternatively, but less desirably, two chromatograms can be run at different wavelengths and compared.

A technique particularly useful for single-channel detection systems, but one which can also be used with diode-array and similar detectors, is to differentiate

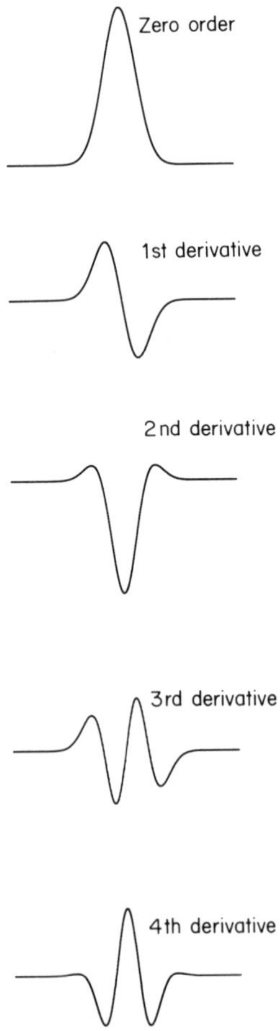

Figure 7.5. Zero-order and first to fourth derivatives of a Gaussian peak. Note the inversion of the second and fourth derivatives

the detector signal with respect to time.[1,7,14,34] Differentiation is simply the determination of the slope of the detector signal at all points. This method forms the basis for peak detection in many integrators.[35] The principal features of the first to fourth derivatives of an ideal Gaussian peak are shown in Figure 7.5. Although the shapes of the peaks become more complex as the order of the derivative increases, it is clear that the width of the main peak is progressively reduced. It is the reduction in band widths which makes derivatives particularly useful in resolution enhancement. It must be emphasized, however, that there can be no resolution enhancement if total peak overlap has occurred. The resolution enhancement attainable in practice will be limited by interference from the satellite peaks, which arise on the sides of the main bands as the derivative order increases, and by the increase in noise level which is inevitable as higher order derivatives are generated. Nevertheless, the second derivative has been shown to be invaluable as an aid to determining peak purity in automated optimization using the simplex procedure[36] (see Chapter 6); a plot of the second derivative of an 'optimized' separation showed an apparently pure peak to consist of at least two components (Figure 6.9). The relative amplitude ratio of the second derivative on the leading and trailing edge of a peak has also been suggested as a criterion for peak purity.[37]

7.2.3. Conclusions

Separation optimization should be approached with appropriate objectives and criteria. Thus it will be generally true that powerful post-chromatographic data manipulation cannot be expected to make up for deficiencies in the initial separation. Nevertheless, there will be situations where improvement of the chromatography is elusive or perhaps impossible. Indeed, there are a finite number of peaks which can be separated on a given column.[38–40] For example, a nominal 5000-plate column will have a theoretical peak capacity[39] of approximately 43 over a k' range of 0–10. This figure is derived from equation 7.4, which expresses the peak capacity (n) as a function of the plate count of the column (N), the retention time of the last eluted peak (t_n) and the column dead time, t_0.

$$n = 1 + \tfrac{1}{4}N^{\frac{1}{2}}\ln\,(t_n/t_0) \qquad (7.4)$$

The peak capacity of 165 assumes Gaussian peaks, a resolution of unity and no dependence of the column plate number on the analysis time or capacity factor. However, on a statistical basis,[38] it is estimated that the chromatogram must be 95% vacant to provide a reasonable probability that any given component will appear as an isolated peak! Another conclusion of the statistical approach is that the number of single-component peaks cannot exceed 18% of the peak capacity of the column even under the most favourable circumstances. The implications of this in practice are that this 5000-plate column may prove inadequate for the complete separation of a mixture containing more

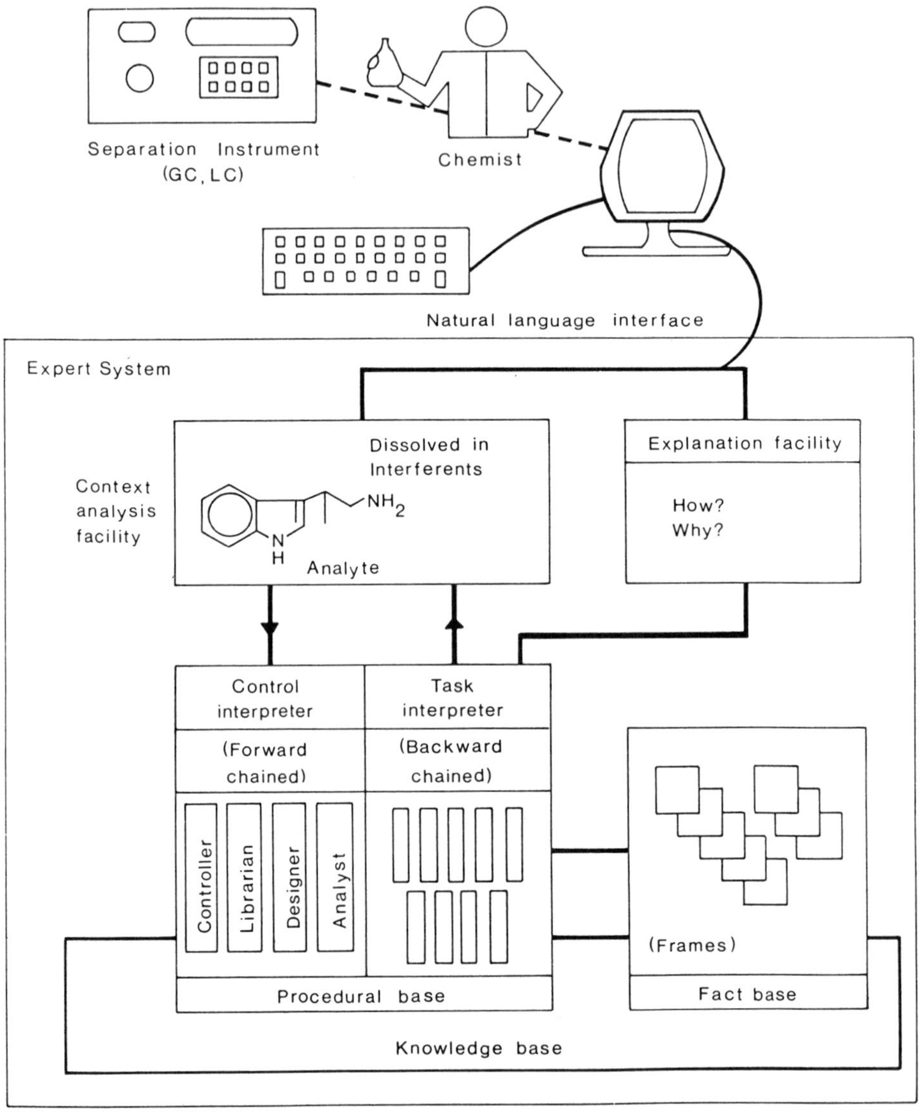

Figure 7.6. Expert system for separation sciences. The user begins by entering the analyte structure(s), sample solvent system and possible interferants using the terminal. The context analysis facility evaluates the problem. The procedural base consists of two parts. The control interpreter contains procedural operations determining the actions of a librarian, an experiment designer, an analyst and the master controller. The task interpreter contains modules reflecting available procedural information on agents and situations. The fact base consists of literature methods and chemical information on analyte compounds. The knowledge base

```
FIRST, ENTER THE GENERIC NAMES OF THE COMPOUNDS YOU WISH TO ANALYZE.
WHEN YOU HAVE NO MORE COMPOUNDS TO ENTER, TYPE NIL
FIRST COMPOUND:   THEOPHYLLINE
NEXT COMPOUND:    DESIPRAMINE
NEXT COMPOUND:    NIL
NEXT, ENTER THE SAMPLE MATRIX:   SERUM
NEXT, ENTER THE INTERFERENCES TO BE RECKONED WITH
COMPOUND:         PLASMA
COMPOUND:         NIL
THANK YOU
NU. OF SOLUTIONS FOUND = 2
I RECOMMEND THAT WE USE THE METHOD IDENTIFIED AS LCAW-87 AS A STARTING
POINT FOR YOUR SEPARATION
DO YOU WANT TO:
    (A)  PROCEED UPON MY RECOMMENDATION
    (B)  CHOOSE A (DIFFERENT) LIBRARY METHOD AS A STARTING POINT
    (C)  STUDY THE LIBRARY METHODS ON YOUR OWN BEFORE DECIDING
>> STUDY
(LCAW-87   (AKO VALUE LIBSEP)))
           (TEXT (VALUE (LCAW) (107)))     /LITERATURE REFERENCE TO METHOD
           (AUTHORS (VALUE (LARSON-L)))
           (TITLE (VALUE (DETERMINATION)
                         (OF)
                         (SOME)
                         (TRICYCLIC)
                         (ANTIDEPRESSANTS)
                         (IN)
                         (SERUM)))
           (ANALYTES (VALUE (DOXEPIN (TR 8.6 MIN))          /INVOLVED ANALYTES
                            (DESIPRAMINE (TR 10.3 MIN))
                            (NORTRYPTYLINE (TR 12.4 MIN))
                            (IMIPRAMINE (TR 13.6 MIN))
                            (AMITRYPTYLINE (TR 16.6 MIN))))
           (INTSTAND (VALUE (P__CHLORODISOPYRAMIDE (TR 6.0 MIN))))
           (NOINT (VALUE (SERUM)))                             /CONDITIONS
           (PROGRAM (VALUE (ISOCRATIC (2.0. MLMIN))))
           (MOBILEA (VALUE (ACN (35%))
                           (KH2PO4 (0.01 MOLAR))
                           (TMACL (0.02 MOLAR))
                           (H3PO4 (1%))))`
           (COLTYPE (VALUE (MCH-10 (MICROPAK))))
           (COLSIZE (VALUE (STANDARD (ID 4.0 MM) (LENGTH 30.0 CM))))
           (DETECTOR (VALUE (UV-ABS (LAMBDA 240 NM) (VARIAN-5020))))
           (SAMPREP (VALUE (SERUM (2.0 ML))                /SAMPLE PREPARATION
                           (MADE-ALKALINE (NAOH))
                           (EXTRACTED (HEXANE ISOAMYL-ALCOHOL))
                           (EXTRACTED2 (H2SO4)))))
```

Figure 7.7. Printout of the interaction between the user and the expert system in Figure 7.6. *Reprinted with permission from Ref. 45, Copyright (1984) American Chemical Society*

---

is made up of the procedural and fact bases. The explanation facility interacts with the user, providing information on how decisions have been reached and why questions have been raised. The system generates the preferred analytical method for an analyte in the specified matrix. *Reprinted with permission from ref. 45. Copyright 1984, American Chemical Society* than 8 components. Therefore, as the mixtures to be separated increase in complexity, any additional information that can be gained from on-line detection or by the application of post-chromatographic methods will prove extremely valuable in aiding the recognition of a truly optimized separation.

7.3. THE FUTURE AND EXPERT SYSTEMS

Unless the decision on the choice of optimization method has been made by the equipment available, one of the great difficulties often experienced when tackling a new separation will be deciding where to start. Given an unknown sample, it can appear almost impossible to know what the first step should be. Some guidance is available[41–43] to indicate certain 'experts'' views on how the problem should be approached. One of the great chemometric advances of recent years is the development of computer software which enables a computer to simulate the abilities of experts, the so-called 'Expert System'. An expert system has been defined[44] as 'An intelligent computer program that uses knowledge and inference procedures to solve problems that are difficult enough to require significant expertise for their solution. The knowledge necessary to perform at such a level plus the inference procedures can be thought of as a model of the expertise of the best practitioner in the field.'

Figure 7.6 shows a representation of a chromatography expert system[45]. The knowledge consists of *facts* and *procedures* (or rules). The *facts* consist of the information that is generally agreed by experts in the field to represent their best conclusions. The *rules*, or heuristics, are rules of reasoning and good judgement which define the processes of good decision making. Expert systems are still in their infancy in analytical chemistry but already there are exciting developments in chromatography. Figure 7.7 shows a printout of a typical interaction between the system and the user. In the example shown the compounds to be analyzed are known and so experimental optimization of the separation has not been necessary. Note the 'friendliness' of the dialogue: this is a key feature of expert systems. Whilst the style may not be to everyone's liking, it is possible to interrogate the system to establish how particular conclusions were arrived at.

It is always dangerous to speculate on future developments. What is certain is that the relatively recent subject of automated separation optimization will receive an ever increasing amount of attention in the years to come. There will be incremental advances in both simultaneous and sequential optimization methods. Chromatographic optimization procedures will become more multidisciplinary as methods are sought to obtain ever more information from the separations. This will be no better epitomized than in the anticipated developments in expert systems. There is little doubt that expert systems will develop to a stage where essentially closed-loop operation is possible. Having presented the system with a totally unknown sample, the results of exploratory experiments will be fed back in in order to provide information which will permit the attainment of the goal of totally automated separation methods development.

7.4. REFERENCES

1. A. F. Fell, *Anal. Proc.*, **17** (1980) 512.
2. S. A. George and A. Maute, *Chromatographia*, **15** (1982) 419.
3. 'Focus', *Anal. Chem.*, **57** (1985) 276A.
4. A. F. Fell, *Anal. Proc.*, **20** (1983) 356.
5. A. F. Fell, H. P. Scott, R. Gill and A. C. Moffat, *Chromatographia*, **16** (1982) 69.
6. A. F. Fell, H. P. Scott, R. Gill and A. C. Moffat, *J. Chromatogr.*, **273** (1983) 3.
7. A. F. Fell, H. P. Scott, R. Gill and A. C. Moffat, *J. Chromatogr.*, **282** (1983) 123.
8. A. F. Fell, *Lab. Pract.*, **34** (3) (1985) 11.
9. B. F. H. Drenth, R. T. Ghijsen and R. A. de Zeeuw, *J. Chromatogr.*, **238** (1982) 113.
10. F. Overzet, R. T. Ghijsen, B. F. H. Drenth and R. A. de Zeeuw, *J. Chromatogr.*, **240** (1982) 190.
11. R. T. Ghijsen, B. F. H. Drenth, F. Overzet and R. A. de Zeeuw, *J. High Resolut. Chromatogr. Chromatogr. Commun.*, **5** (1982) 192.
12. F. Overzet, *J. High Resolut. Chromatogr. Chromatogr. Commun.*, **5** (1982) 604.
13. F. Overzet, A. Rurak, H. van der Voet, B. F. H. Dreuth, R. T. Ghijsen and R. A. de Zeeuw, *J. Chromatogr.*, **267** (1983) 329.
14. B. J. Clark, A. F. Fell, H. P. Scott and D. Westerlund, *J. Chromatogr.*, **286** (1984) 271.
15. Using a Hewlett-Packard 85B computer and Model 1040A detector, Figure 7.2 took 2 h to produce.
16. A. Fell, B. J. Clark and H. P. Scott, *J. Pharm. Biomed. Anal.*, **1** (1983) 557.
17. C. J. Warwick and D. A. Bagon, *Chromatographia*, **15** (1982) 443.
18. G. T. Carter, R. E. Schiesswohl, H. Burke and R. Yang, *J. Pharm. Sci.*, **71** (1982) 317.
19. A. C. J. H. Drouen, H. A. H. Billiet and L. de Galan, *Anal. Chem.*, **56** (1984) 971.
20. K. Zech, R. Huber and H. Elgass, *J. Chromatogr.*, **282** (1983) 161.
21. T. Hoshino, M. Senda, T. Hondo, M. Saito and S. Tohei, *J. Chromatogr.*, **316** (1984) 473.
22. E. R. Malinowski and D. G. Howery, *Factor Analysis in Chemistry*, Wiley, New York, 1981.
23. J. Kruskal, 'Factor Analysis and Principal Components Analysis, the Bilinear Methods,' in *Encyclopedia of Statistics*, The Free Press, New York, 1978.
24. J. M. Halket, *J. High Resolut. Chromatogr. Chromatogr. Commun.*, **2** (1979) 197.
25. M. McCue and E. R. Malinowski, *J. Chromatogr. Sci.*, **21** (1983) 229.
26. A. Lorber, *Anal. Chim. Acta.*, **164** (1984) 293.
27. D. W. Osten and B. R. Kowalski, *Anal. Chem.*, **56** (1984) 991.
28. A. C. J. H. Drouen, H. A. H. Billiet and L. de Galan, *Anal. Chem.*, **57** (1985) 962.
29. M. Rosenbaum, V. Hancil and R. Komers, *J. Chromatogr.*, **191** (1980) 157.
30. R. A. Vaidya and R. D. Hester, *J. Chromatogr.*, **287** (1984) 231.
31. S. T. Balke, *Quantitative Column Liquid Chromatography, a Survey of Chemometric Methods*, Elsevier, Amsterdam, 1984.
32. R. Yost, W. MacLean and J. Stoveken, *Chromatogr. Newsl.*, **4** (1976) 1.
33. P. C. White, *J. Chromatogr.*, **200** (1980) 271.
34. J. C. Berridge and K. S. Andrews, *Analyst (London)*, **109** (1984) 287.
35. J. L. Excoffier and G. Guiochon, *Chromatographia*, **15** (1981) 543.
36. J. C. Berridge, *Chromatographia*, **16** (1982) 172.
37. E. Grushka and G. C. Monacelli, *Anal. Chem.*, **44** (1972) 484.
38. J. M. Davis and J. C. Giddings, *Anal. Chem.*, **55** (1983) 418.
39. E. Grushka, *J. Chromatogr.*, **316** (1984) 81.
40. M. Martin and G. Guiochon, *Anal. Chem.*, **57** (1985) 289.

41. P. J. Schoenmakers, H. A. H. Billiet and L. de Galan, *J. Chromatogr.*, **205** (1981) 13.
42. P. J. Schoenmakers, *Chem. Mag.*, February (1983) 68.
43. P. J. Colley, Paper presented at the Pittsburg Conference on Analytical Chemistry and Applied Spectroscopy, Atlantic City, NJ, 1982, Paper 799.
44. E. A. Feigenbaum, in D. Michie (Ed.), *Expert Systems in the Microelectronic Age*, Edinburgh University Press, Edinburgh, 1979.
45. J. Karnicky, *Anal. Chem.*, **56** (1984) 1312A.

Appendix 1. Equipment Details

APPARATUS

System 1

Hewlett-Packard 1090A, consisting of the following components and conditions:

Pump	DR5 ternary pumping system.
Injector	Autosampler with Rheodyne 7017 valve.
Column	6 cm × 4.6 mm i.d. containing Hypersil ODS (spherical silica with octadecyl-bonded stationary phase, nominal diameter 3 μm).
Column temperature	40°C.
Detector	Integral 1040A diode-array.
Wavelength	260 ± 40 nm.
Integrator	Integral DPU (8-channel data processing unit).
Plotter	HP 7470A.
System control	HP 85B microcomputer with 5 MB Winchester disc.

System 2

Laboratory Data Control Analyst 7800, consisting of the following components and conditions:

Pumps	Three ConstaMetric IIG pumps.
Injector	Autosampler (Varian) with Rheodyne 7010 valve.
Column	6 cm × 4.6 mm i.d. containing Hypersil ODS (spherical silica with octadecyl-bonded stationary phase, nominal diameter 3 μm (as for System 1).
Column temperature	Ambient (20–22°C).
Detector	SpectroMonitor D.

Wavelength	260 nm.
System control	Chromatography Control Module: includes Rev. C firmware for control of two pumps and the detector and integration (the third pump was controlled through a spare analogue port addressed directly by BASIC commands)
	BASIC interpreter.
	32k RAM.
	2 MB floppy disc (Commodore 8250).

All experiments were carried out using System 1 except for the fully automated optimization using the sequential simplex procedure described in Chapter 5.

REAGENTS AND SOLUTES

All solvents were of HPLC grade (Rathburn Chemicals, Peebles, Scotland) except acetonitrile, which was of far-UV grade. Solutes were of analytical-reagent grade and were used as received.

Solute Identities

Number	*Identity*
1	3-Hydroxybenzaldehyde
2	Phenol
3	Propyl-*p*-hydroxybenzoic acid
4	Butyl-*p*-hydroxybenzoic acid
5	Diethyl phthalate
6	Toluene

Appendix II. BASIC Program for the Translation of Gradient Elution Retention Time to Isocratic Composition

```
100 CLS
110 PRINT "GRADIENT ELUTION TO ISOCRATIC COMPOSITION"
120 PRINT "TRANSLATION ACCORDING TO METHOD OF SNYDER ET AL"
130 PRINT "BY J C BERRIDGE     JANUARY 1985"
140 INPUT "DO YOU WANT INSTRUCTIONS (Y/N)";A$
150 IF LEFT$(A$,1) = "Y" THEN CLS: GOSUB 5000

1000 CLS: REM  MAIN PROGRAM
1010 INPUT "ENTER COLUMN VOID TIME (MIN) ";T0
1020 INPUT  "ENTER  CHROMATOGRAPHIC  VOID  TIME  [MIN:- =
     VOL(ML)/FLOW (ML/MIN)]";TD
1030 INPUT "ENTER RATE OF METHANOL CHANGE (% PER MIN)";F2
1040 INPUT "ENTER PEAK RETENTION TIME (MIN)";RT
1050 INPUT "ENTER DESIRED ISOCRATIC RETENTION TIME ";TNEW

1100 REM  CALCULATE EXIT COMPOSITION AT RT
1110 EXIT = INT(F2*(RT - T0 - TD)*100+.5)/100
1120 IF EXIT<=0 THEN EXIT = .001

1200 REM CALCULATE WATER CAPACITY FACTOR
1210 K0 = 2.2*10^(.2*(RT-T0)/T0)-2.2

1300 REM  CALCULATE K(FINAL)
1310 KF = 1/(2.303*.2+(1/K0))

1400 REM   S1 = SOLVENT STRENGTH
1410 S1 = (LOG(K0)-LOG(KF))/(2.303*EXIT*.01)

1500 REM   KNEW IS NEW CAPACITY FACTOR FOR ISOCRATIC
1510 KNEW = (TNEW -T0)/T0 : IF KNEW <=0 THEN 1800
1520 MEOH = (LOG(K0)-LOG(KNEW))/(S1*2.303):IF MEOH<0  THEN
     PRINT  "THIS  RETENTION TIME MAY NOT BE  POSSIBLE  USE
```

```
     100% WATER OR TRY A DIFFERENT CHROMATOGRAPHIC MODE":
     GOTO 1640
1530 IF MEOH>1 THEN 1800
1540 MECN = MEOH * MEOH * .32 + MEOH * .57
1550 THF = MEOH * .66

1600 PRINT "% METHANOL = "MEOH*100
1610 PRINT "% MECN =     "MECN*100
1620 PRINT "%THF =       "THF*100
1630 PRINT

1640 INPUT "INVESTIGATE ANOTHER RETENTION TIME FOR AN
     ISOCRATIC PEAK";A$
1650 IF LEFT$(A$,1)="Y" THEN 1050

1660 INPUT "INVESTIGATE ANOTHER PEAK FROM THE GRADIENT";A$
1670 IF LEFT$(A$,1)="Y" THEN 1040
1700 END

1800 PRINT "SORRY, THIS RETENTION IS NOT POSSIBLE"
1810 GOTO 1640

5000 REM INSTRUCTIONS
5010 PRINT "THIS PROGRAM WILL ALLOW THE PREDICTION OF THE
     REQUIRED ISOCRATIC REVERSED -PHASE CONDITIONS FOR A
     PEAK ELUTED UNDER GRADIENT ELUTION"
5020 PRINT "THE GRADIENT ELUTION MUST BE FOR A LINEAR
     GRADIENT CARRIED OUT WITH A RATE OF CHANGE OF
     (6.67/T0)% METHANOL/MIN"
5030 PRINT "THE DATA REQUIRED ARE THE COLUMN VOID TIME IN
     MINUTES"
5040 PRINT "     THE DELAY TIME CORRESPONDING TO THE VOLUME OF"
5050 PRINT "     TUBING BETWEEN THE PUMP/MIXERS AND THE COLUMN"
5060 PRINT "AND THE PEAK RETENTION TIMES"
5070 PRINT
5080 INPUT "PRESS RETURN TO CONTINUE ";A$
5090 RETURN
```

Appendix III. BASIC Program for Pragmatic Optimization of Mobile Phase Composition and Temperature

```
100 CLS
110 PRINT "LINEAR OPTIMISATION PROGRAM"
120 PRINT "VERSION 2.1. J. C. BERRIDGE   JANUARY 1985"
130 INPUT "DO YOU WANT INSTRUCTIONS (Y/N)";A$
140 IF LEFT$(A$,1) = "Y" THEN GOTO 5000
150 REM DIMENSION ARRAYS AND SET VARIABLES
160 DIM A(5,5), J(5,5), K(5,5): X=1
190 CLS

200 REM INPUT ALL COLUMN DETAILS
210 INPUT "COLUMN LENGTH (MM)";L
220 INPUT "PARTICLE DIAMETER (MICRON)";M(1)
230 INPUT "COLUMN RESISTANCE FACTOR - IF UNKNOWN  ENTER
    1000";P5
240 INPUT "ORGANIC SOLVENT: METHANOL = 1, ACETONITRILE =
    2";M
250 INPUT "WHAT STANDARD TEMPERATURE WAS USED (DEG C)";T
260 INPUT "STANDARD ORGANIC COMPOSITION (% MODIFIER)";F1
270 INPUT  "APPROXIMATE  MOLECULAR  WEIGHTS  OF  THE  2
    SOLUTES- SOLUTE 1";G1
280 INPUT "........... AND SOLUTE 2";G2
290 F1=F1/100:T1 = T1+273: REM USE ABSOLUTE TEMPERATURE
300 INPUT "COLUMN VOID TIME (SECS)";R1
310 PRINT "ENTER THE CAPACITY FACTORS OF THE 2 PEAKS AT"
    T
315 PRINT "DEG FOR THE STANDARD COMPOSITION"
320 FOR I = 1 TO 2
330    PRINT "FOR PEAK NO. "I
340    INPUT "ENTER CAPACITY FACTOR";K(X,I)
350 NEXT I
360 CLS
370 IF X=2 THEN GOTO 400
```

```
380 INPUT  "WHAT WAS THE OTHER ORGANIC  COMPOSITION  (%
    MODIFIER)";F2
390 F2=F2/100
400 PRINT "NOW ENTER THE CAPACITY FACTORS AT "T" DEGREES
    FOR " F2*100" % ORGANIC"
410 FOR I = 1 TO 2
420    PRINT "FOR PEAK NO. "I
430    INPUT "ENTER CAPACITY FACTOR "; J(X,I)
440 NEXT I
450 IF X=2 THEN GOTO 500
460 X=2
470 INPUT "WHAT OTHER TEMPERATURE WAS USED (DEG C)";T
480 T2 = T+273
490 GOTO 310

500 REM NOW CALCULATE COEFFICIENT C
510 FOR I=1 TO 2
520    C(I)=LOG(J(1,I))-LOG(K(1,I))
530    C(I)=ABS(C(I)/2.303):C(I)=C(I)/(F2-F1)
540 NEXT I

600 REM CALCULATE VALUES OF COEFFICIENT A
610 FOR I = 1 TO 2
620   A(1,I)=(LOG(K(1,I))-LOG(K(2,I)))/((1/T1)-(1/T2))
630   A(I,I)=ABS(A(1,I)/2.303)

650 REM SECOND COMPOSITION
660   A(2,I) = (LOG(J(1,I))-LOG(J(2,I)))/((1/T1)-(1/T2))
670   A(2,I) = ABS(A(2,I)/2.303)
680   A(I,0)=(A(1,I)+A(2,I))/2

700 REM GALCULATE GRADIENT COEFFICIENT E
710    E(I)= (A(2,I)-A(1,I))/(F2-F1)
720    E(I)=ABS(E(I))

800 REM CALCULATE INTERCEPT COEFFICIENT D
810    D(I)= A(I,0)+E(I)*((F2+F1)/2)
820    D(I) = ABS(D(I))

900 REM CALCULATE FACTOR B
910    B(I) = ((LOG(K(1,I))/2.303)+C(I)*F1)
920    B(I)=(B(I)+((LOG(J(1,I))/2.303)+C(I)*F2))/2
930    B(I)=ABS(B(I))
940 NEXT I

1000 REM MAIN PROGRAM FOR PREDICTION OF RETENTION TIMES

1010 REM CALCULATE K AS A FUNCTION OF T
1020 CLS
1030 INPUT "WHAT TEMPERATURE DO YOU WANT TO INVESTIGATE
     (DEG C)";T
1040 T3=T+273
1050 INPUT "WHAT ORGANIC PERCENTAGE";F
1060 F3=F/100

1100 REM CALCULATE NEW CAPACITY FACTORS
1110 FOR I=1 TO 2
```

```
1120     L(I)=B(I)-F3*C(I)
1130     K(I,0)=L(I)-(A(I,0)*((1/T1)-(1/T3)))
1140     K(I,0)=EXP(K(I,0))
1150 NEXT I

1200 REM FIND K MAX
1210 IF K(1,0)<K(2,0) THEN 1300
1220 K=K(1,0):K(1,0)=K(2,0):K(2,0)=K

1300 REM LINEAR VELOCITY   U1
1310 U1 = L/R1

1330 REM  CALCULATE MOBILE PHASE VISCOSITY
1340 IF M=1 THEN GOSUB 2000:  REM  MUST BE METHANOL
1350 IF M=1 THEN GOTO 1370
1360 GOSUB 2200
1370 GOSUB 2400

1400 REM CALCULATE REDUCED VELOCITY
1410 U2=U1*(1E-09)*M(1)/Y

1420 REM REDUCED PLATE HEIGHT      H1
1430 H1=U2^.33+.1*U2+(2/U2):O=1

1440 REM CALCULATE NO OF PLATES     P1 - CAN BE  MODIFIED
     BY FACTOR O
1450 P=INT(L*1000/(H1*M(1)))
1460 P1=P*O
1470 CLS

1480 PRINT  "THE  NUMBER OF PLATES FOR  THE  COLUMN  IS
     ESTIMATED TO BE "P1
1490 PRINT "THIS ASSUMES PERFECT GAUSSIAN PEAKS AND MAY
     BE AN OVER-ESTIMATE"
1500 INPUT "IF THE VALUE IS TOO HIGH,  PLEASE  ENTER  A
     SCALING  FACTOR  BETWEEN  0  AND 1  (1  =  CONTINUE
     UNCHANGED)";O
1510 IF O=1 THEN 1600
1520 IF O<1 THEN 1460
1530 GOTO 1470

1600 REM RETENTION TIME      R9
1610 R9=R1*(1+K(1,0))
1620 R=R1*(1+K(2,0))

1630 REM   PEAK WIDTHS
1640 W1 =4*R9/SQR(P1):W2 = 4*R/SQR(P1)

1650 REM  RESOLUTION    R5
1660 R5 = 2*(R-R9)/(W1+W2)

1670 REM   INLET PRESSURE   P9
1680 P9 = (U1*P5*V1*L*.146)/(M(1)*M(1))

1700 REM   OUTPUT
1710 CLS
```

```
1720 K(1,0)=INT(K(1,0)*100+.5)/100:K(2,0)  =
     INT(K(2,0)*100+.5)/100
1730 P9=INT(P9):R9=INT(R9):R5=INT(R5*10+.5)/10
1740 W1=INT(W1*10+.5)/10:W2=INT(W2*10+.5)/10
1750 PRINT TAB(24);"PEAK 1";TAB(34);"PEAK 2"
1760 PRINT "CAPACITY FACTOR"  TAB(24);  K(1,0);TAB(34);
     K(2,0)
1770 PRINT "RETENTION TIME (SEC)" TAB(24);  R9; TAB(34);
     INT(R)
1780 PRINT "BASE WIDTH (SEC)" TAB(24);W1;TAB(34);W2
1790 PRINT
1800 PRINT "RESOLUTION"TAB(30);R5
1810 PRINT "INLET PRESSURE (POUNDS/SQ IN)"TAB(30) P9
1820 PRINT "MOBILE PHASE VISCOSITY (CPS)" TAB(30) V1
1830 PRINT
1840 INPUT "INVESTIGATE ANOTHER CONDITION (Y/N)";A$
1850 IF LEFT$(A$,1)="Y" THEN 1000
1900 END

2000 REM   METHANOL VISCOSITY
2010 V2=1.9:W=32

2020 REM    VISCOSITY AT 40 DEGREES
2030 V3=.64+(2.1*F3)-2.32*F3*F3
2040 T4=(1/T3)-(1/313)
2050 V1=(LOG(V3)/2.303)+1000*T4*(.7+(.2*V3))
2060 V1=INT((10^V1)*100+.5)/100
2070 RETURN


2200 REM    ACETONITRILE VISCOSITY
2210 V2=1.9:W=41

2220 REM   V3= VISCOSITY AT 40 DEGREES
2230 V3=.26+.78*(1-F3)-.374*(1-F3)^3
2240 T4=(1/T3)-(1/313)
2250 V1=(LOG(V3)/2.303)+1000*T4*(.7+(.2*V3))
2260 V1=INT((10^V1)*100+.5)/100
2270 RETURN

2400 REM   WILKIE-CHANG CALCULATION
2410 Z=(F3*W*V2)+((1-F3)*2.6*18)

2420 REM   DIFFUSION COEFFICIENT   Y
2430 G = ((G1+G2)/2)^.6:Y =7.4E-12*SQR(Z)*T3/(V1*G)
2440 RETURN

5000 CLS
5010 PRINT "THIS PROGRAM PREDICTS THE RETENTION TIMES OF
     HPLC PEAKS"
5020 PRINT "IN REVERSED PHASE CHROMATOGRAPHY FOR  MOBILE
     PHASES CONTAINING"
5030 PRINT "METHANOL OR ACETONITRILE AS MODIFIERS"
5040 PRINT "TWO PEAKS CAN BE INVESTIGATED AND THE  DATA
     REQUIRED ARE:-"
5050 PRINT "THE CAPACITY FACTORS OF THESE TWO PEAKS AT 2
     TEMPERATURES"
```

```
5060 PRINT "AND TWO COMPOSITIONS;  I.E. FOUR EXPERIMENTS
     IN ALL"
5070 PRINT "ALSO NEEDED ARE THE COLUMN VOID TIME AND THE
     STANDARD DATA "
5080 PRINT "ABOUT THE COLUMN AND THE PACKING"
5090 PRINT "FOR MORE INFORMATION SEE SNYDER ET  AL,  J.
     CHROM 185 (1979) 153"
5100 PRINT "AND BERRIDGE J. CHROM 202 (1980) 469"
5200 INPUT "PRESS RETURN TO CONTINUE ";A$
5300 GOTO 150
```

Appendix IV. Simple BASIC Program for Triangular Mixture Design Plot

```
10  REM SIMPLE TRIANGULAR PLOT PROGRAM FOR OVERLAPPING RESOLUTIC
    MAPS
20 REM  BY J C BERRIDGE  JANUARY 1985
30 REM  WRITTEN IN MICROSOFT BASIC FOR EPSON QX-10
40 REM ***********************************************************
50 REM  INITIALISE VARIABLES AND DIMENSION ARRAYS
60 DIM VARIABLE$(4),RESULT(10),COEFF(7)
70 Y$(0)=".":Y$(1)="-":Y$(2)="+":Y$(3)="*":Y$(4)="#"


100 CLS: REM CLEAR SCREEN
110 PRINT "TRIANGULAR PLOT FOR SIMPLEX LATTICE  DESIGNS":  PRINT
    PRINT
120 INPUT  "ENTER TITLE FOR PLOT"; HEADER$
130 INPUT "ENTER NAME OF FIRST VARIABLE "; VARIABLE$(1)
140 INPUT "ENTER NAME OF SECOND VARIABLE "; VARIABLE$(2)
150 INPUT "ENTER NAME OF THIRD VARIABLE "; VARIABLE$(3)
160 INPUT "ENTER NAME OF MEASURED RESPONSE "; VARIABLE$(4)

170 CLS: REM RESULTS ENTRY
180  PRINT  "THE  RESULTS OF AT LEAST SEVEN  EXPERIMENTS  CARRIE
     ACCORDING TO"
190 PRINT "A MIXTURE DESIGN MUST NOW BE ENTERED"
200 PRINT
210 PRINT "RESULT FOR 100% ";VARIABLE$(1);:INPUT RESULT(1)
220 PRINT "RESULT FOR 100% ";VARIABLE$(2);:INPUT RESULT(2)
230 PRINT "RESULT FOR 100% ";VARIABLE$(3);:INPUT RESULT(3)
240 PRINT "RESULT FOR 50% ";VARIABLE$(1); " & 50% ";VARIABLE$(2)
    :INPUT RESULT(4)
250 PRINT "RESULT FOR 50% "; VARIABLE$(1);" & 50% ";VARIABLE$(3)
    :INPUT RESULT(5)
260 PRINT "RESULT FOR 50% "; VARIABLE$(2);" & 50% ";VARIABLE$(3)
    :INPUT RESULT(6)
270 PRINT "RESULT FOR 33% ";VARIABLE$(1);" & 33%  ";VARIABLE$(2)
    "  & 33% ";VARIABLE$(3);:INPUT RESULT(7)
```

```
500 REM CALCULATE POLYNOMIAL COEFFICIENTS
510 FOR I = 1 TO 3
520    COEFF(I) = RESULT(I)
530 NEXT I
540 COEFF(4)  =  4*RESULT(4)-2*(RESULT(1)+RESULT(2))
550 COEFF(5)  =  4*RESULT(5)-2*(RESULT(1)+RESULT(3))
560 COEFF(6)  =  4*RESULT(6)-2*(RESULT(2)+RESULT(3))
570 COEFF(7)  =  27 *  RESULT(7)-12*(RESULT(4)+RESULT(5)+RESULT(6))
    + 3 *(RESULT(1)+RESULT(2)+RESULT(3))


1000 CLS: REM LACK OF FIT CALCULATION
1010  INPUT "DO YOU WANT TO ENTER MORE POINTS TO CHECK FIT  (Y/N)
     ";A$

1020 IF LEFT$(A$,1) <> "Y" THEN 2000
1030 INPUT "ENTER NUMBER OF POINTS TO CHECK FIT ";POINTS
1040 IF POINTS<=0 THEN 1030
1050 LACK=0
1060    FOR I=1 TO POINTS
1070    PRINT "FOR EXPERIMENT NO "I;
1080      FOR J= 1 TO 2
1090      PRINT "ENTER DECIMAL FRACTION OF  ";VARIABLE$(J);:INPUT
          DECFR(J)
1100      NEXT J
1110    DECFR(3)=1-DECFR(2)-DECFR(1)
1120    INPUT "ENTER EXPERIMENTAL RESULT ";EXPVAL

1130    GOSUB 5000
1140    DIFF = (EXPVAL-CALVAL)*(EXPVAL-CALVAL)
1150    LACK= LACK+DIFF
1160    NEXT I
1170 FITVAR=LACK/POINTS


2000 CLS:REM PLOTTING VARIABLES
2010 GOSUB 4000
2020 PRINT "MAX VALUE OF RESPONSE IS ";MAX
2030 PRINT "MIN VALUE OF RESPONSE IS ";MIN
2040  INPUT  "ENTER NUMBER OF LEVELS TO BE  PLOTTED  (MAXIMUM  4)
     ";LEVELS
2050 IF LEVELS>4 THEN 2040
2060    FOR I=1 TO LEVELS
2070    PRINT  "FOR  LEVEL ";I;:INPUT "ENTER  MAXIMUM  VALUE   ";
        LEVELS(I)
2080    NEXT I
2090 IF LEVELS <5 THEN FOR I=LEVELS+1 TO 5:LEVELS(I)=1E+29:NEXT I


3000 REM                 PRINT AND PLOT RESULTS AND SURFACE
3010 REM                 FOR AN 80 COLUMN PRINTER (EPSON FX-80)

3020 LPRINT "  PLOT OF ";HEADER$:LPRINT
3030 FOR I=1 TO 12 + LEN(HEADER$):LPRINT"=";:NEXT I:LPRINT:LPRINT
3040 LPRINT "COEFFICIENTS OF POLYNOMIAL Y = AX+BY+CZ+DXY + EXZ  +
     FYZ+GXYZ ARE..."
3050 LPRINT
```

```
3060    FOR I=1 TO 7
3070    PRINT "COEFFICIENT "I,COEFF(I)
3080    LPRINT "COEFFICIENT "I,COEFF(I)
3090    NEXT I
3100  LPRINT:IF  POINTS>0  THEN LPRINT "LACK OF  FIT  VARIANCE
     ";FITVAR:LPRINT
3110 LPRINT "RESPONSE IS ";VARIABLE$(4):LPRINT

3120 FOR I=1 TO LEVELS: LPRINT "PLOT LEVEL UP TO  ";LEVELS(I),"
     ";Y$(I-1) :NEXT I
3130  LPRINT  "PLOT  OVER LEVEL    ";LEVELS(LEVELS),"  =         "
      Y$(LEVELS):LPRINT
3140  LPRINT  "MAXIMUM  RESPONSE IS ";MAX;" AT "  COORDA*100  "%
     "COORDB * 100 "%: " (1-COORDA-COORDB)*100"%"

3150 LPRINT:LPRINT:LPRINT TAB(39);"0"
3160    FOR C=100 TO 0 STEP -5
3170    IF C=70 THEN LPRINT "VARIABLE: ";VARIABLE$(1);
3180    TB=20+C/5
3190    TC=TB-3-INT(LOG(106-C)/2.3)
3200    IF C>1 THEN LPRINT TAB(TC);105-C;
3210    LPRINT TAB(TB);
3220      FOR B=0 TO 100-C STEP 5
3230       DECFR(2)  = B/100:  DECFR(3) = C/100:  DECFR(1)  =  1
           DECFR(2) - DECFR(3)
3240       GOSUB 5000
3250       PLOT$=Y$(0)
3260          FOR  I=1  TO LEVELS:  IF CALVAL  >  LEVELS(I)  THE
              PLOT$ = Y$(I)
3270          NEXT I
3280      LPRINT PLOT$;" ";
3290      NEXT B
3300    LPRINT C:NEXT C
3310  LPRINT TAB(TB);"0";:LPRINT  TAB(TB+20);VARIABLE$(2);:LPRIN
      TAB(TB+39);"100%"

3320 STOP


4000 REM        FIND MAX AND MIN RESPONSES
4010 MIN=1E+29: MAX=-1E+29
4020    FOR A= 0 TO 1.05 STEP .05
4030      FOR B=0 TO 1.05-A STEP .05
4040      DECFR(1)=A:DECFR(2)=B:DECFR(3)=1-A-B
4050      GOSUB 5000
4060       IF  CALVAL>MAX THEN MAX =CALVAL :  COORDA =  DECFR(1)
          COORDB =  DECFR(2)
4070      IF CALVAL<MIN THEN MIN =CALVAL
4080 NEXT B:NEXT A
4090 RETURN


5000 REM  SUBROUTINE TO CALCULATE RESPONSE
5010 CALVAL=COEFF(1)*DECFR(1)+COEFF(2)*DECFR(2)+COEFF(3)*DECFR(3
     +  COEFF  (4)  * DECFR(1)  * DECFR(2)  + COEFF(5)  *  DECFR(1)
     DECFR(3)    +  COEFF(6)*DECFR(2)   *   DECFR(3)   +  COEFF(7)
     DECFR(1)        * DECFR(2)  * DECFR(3)
5020 RETURN
```

Appendix V. BASIC Program for Variable Step-size Simplex Procedure

```
1000 CLS: PRINT "VARIABLE SIZE SIMPLEX OPTIMISATION"
1010 PRINT "VERSION 4.4 BY J C BERRIDGE   JANUARY 1985 "
1020 INPUT "DO YOU WANT INSTRUCTIONS";A$
1030 IF LEFT$(A$,1)="Y" THEN GOSUB 8000

1500 REM DEFINE VARIABLES
1510 DIM P(13,11):REM ARRAY FOR VERTICES
1520 B1 =1: REM SIMPLEX STATUS

2000 REM  MAIN PROGRAM START
2010 CLS
2020 INPUT "HOW MANY VARIABLES IN THE SIMPLEX (2 -9) " ;
     VAR
2030 IF VAR>1 AND VAR <10 THEN 2040 ELSE 2020
2040 INPUT "ENTER OPTIMUM RESPONSE HOPED FOR";M
2050 W6=M: REM   W6 USED TO TRACK BEST VERTEX

2100 REM INPUT ARRAY OF FIRST EXPERIMENTS AND RESULTS
2110 CLS
2120 PRINT  "NOW  ENTER THE  EXPERIMENTAL  DETAILS  AND
     RESULTS FOR THE FIRST ";VAR+1;" EXPERIMENTS"
2150 FOR I=1 TO VAR+1
2160   PRINT
2170   FOR J=1 TO VAR
2180     PRINT "ENTER VARIABLE NUMBER ";J;
2190     INPUT P(I,J)
2200   NEXT J
2210   PRINT
2220   INPUT  "ENTER  RESULT  FOR  THIS  EXPERIMENT  ";
       P(I,VAR+1)
2230   IF P(I,VAR+1)>P(I-1,VAR+1) THEN W = M - P(I,VAR+1)
2240 NEXT I

3000 REM  START OF SIMPLEX ALGORITHM
3010 B=1: REM B IS SIMPLEX EXPANSION COEFFICIENT
```

```
3020 GOSUB 6000
3030 W=W2

3040 GOSUB 7000

4000 REM  CHECK WHETHER BEST VERTEX HAS BEEN RETAINED AGAIN
4010 IF W8<VAR+1 THEN 4500
4020 CLS
4030 PRINT "THE BEST VERTEX HAS BEEN RETAINED IN "VAR+1"
     EXPERIMENTS AND MAY BE A FALSELY HIGH RESULT."
4040 PRINT  "IT IS SUGGESTED THAT YOU YOU  RECHECK  THE
     RESULT OF YOUR BEST EXPERIMENT SO FAR"
4050 I=W9: W8 = 0
4060 FOR J=1 TO VAR
4070   PRINT "FOR VARIABLE NUMBER "J;"     ";P(I,J)
4080 NEXT J
4090 INPUT "PRESS RETURN TO CONTINUE";A$

4500 REM   PRINT NEW VARIABLES
4510 CLS
4520 PRINT "THE NEXT EXPERIMENT SHOULD USE "
4530 I=VAR+2
4540 FOR J=1 TO VAR
4550   PRINT "FOR VARIABLE NUMBER ";J"; : ":P(I,J)
4560 NEXT J
4600 IF B1=3 THEN B1=1
4610 P(VAR+2,VAR+1) = 1E+20
4700 PRINT
4710 INPUT "ARE THESE VALID CONDITIONS (Y/N)";A$
4720 IF LEFT$(A$,1)<>"N" THEN 4800
4730 PRINT "ENTER A VERY POOR RESULT"
4740 W8=W8-1: REM  DON'T COUNT THIS AS AN EXPERIMENT

4800 REM  INPUT NEXT RESULTS
4810 I=VAR+2:B=1
4820 PRINT
4830 INPUT "ENTER THE RESULT ";P(I,VAR+1)
4840 DIST = ABS(M-P(VAR+2,VAR+1))

4850 REM      W6 IS BEST RESULT SO FAR
4860 IF DIST < W6 THEN W6 = DIST: W8 =0

4870 REM  CHECK WHETHER BEST VERTEX RETAINED AGAIN
4880 IF DIST >= W3 THEN W8 = W8+1
4890 IF B1=5 THEN GOTO  5000
4900 IF DIST>= W3 AND DIST <W4 AND B1=1 THEN GOTO 5010
4910 IF DIST >= W5 AND B1=1 THEN B=-.5
4920 IF DIST <W5 AND DIST > W4 AND B1=1 THEN B=.5
4930 IF B<1 THEN B1=5
4940 IF DIST <W3 THEN B1=B1+1
4950 IF DIST < W3 AND B1=2 THEN B=2
4960 IF DIST <W3 AND B1=3 THEN GOTO 5000
4970 IF B1=2 AND B=1 THEN 6500
4980 GOTO 3040
5000 B1=1
5010 FOR J=1 TO VAR+1
5020   P(W,J) = P(VAR+2,J)
5030 NEXT J
5040 GOTO 3020
```

```
6000 REM   FIND THE WORST VERTEX
6010 W2=0
6020 FOR K = 1 TO 2
6030   X1=1: W1=0 : W3 = 1E+20: J = VAR + 1
6040   FOR I=1 TO VAR + 1
6050     IF I = W2 THEN GOTO 6120
6060     X = M - P(I,J)
6070     IF ABS(X)>X1 THEN W1 = I: X1 = ABS(X)

6100     REM    W9 IS POINTER TO BEST VERTEX
6110     IF ABS(X) <=W3 THEN W3= ABS(X): W9 = I
6120   NEXT I
6130   IF W2=0 THEN W5 = X1: W2 = W1
6140   W4 = X1
6150 NEXT K
6160 RETURN

6500 REM    USE LAST VERTEX
6510 FOR J = 1 TO VAR + 1
6520   P(W,J) = P(VAR+3,J)
6530 NEXT J
6540 B1=1: W9 = W
6550 GOTO 3020

7000 REM   REFLECTED VERTEX
7010 IF B<>2 THEN 7050
7020 FOR J=1 TO VAR + 1
7030   P(VAR+3,J) = P(VAR+2,J)
7040 NEXT J
7050 FOR J = 1 TO VAR
7060   P0=0
7070   FOR I = 1 TO VAR + 1
7080     IF I = W THEN GOTO 7100
7090     P0 = P0 + P(I,J)
7100   NEXT I
7110   P0 = P0/VAR: P(VAR+2,J) = P0 + B*(P0 - P(W,J))
7120 NEXT J
7140 RETURN

8000 REM   INSTRUCTIONS FOR USE
8010 CLS
8020 PRINT "THIS PROGRAM EMPLOYS THE VARIABLE STEP  SIZE
     SIMPLEX  ALGORITHM OF NELDER AND MEAD FOR  FUNCTION
     MAXIMISATION"
8030 PRINT "THE PROGRAM WILL HANDLE 2 TO 9 VARIABLES FOR
     SIMULTANEOUS OPTIMISATION"
8040 PRINT "IF YOU HAVE 'N' VARIABLES, 'N+1' EXPERIMENTS
     MUST  BE FIRST CARRIED OUT TO DESCRIBE THE  INITIAL
     SIMPLEX"
8050 PRINT  "IF  THE  COORDINATE  FOR  A  PARTICULAR
     EXPERIMENT LIES OUTSIDE THE NORMAL  CONDITIONS,  DO
     NOT  CARRY OUT AN EXPERIMENT BUT INSTEAD  ASSIGN  A
     VERY UNDESIREABLE RESULT"
8060 PRINT  "THIS  PROCEDURE  ALSO  MONITORS  THE  BEST
     RESPONSE  AND  WILL  WARN  YOU  TO  REPEAT  A
     DETERMINATION  IF  IT SEEMS THAT A GOOD  RESULT  IS
     OVER-INFLUENCING THE OPTIMISATION"
8070 INPUT "PRESS RETURN TO CONTINUE ";A$:RETURN
9000 END
```

Index